"The Trias of energy, development and povert
understood together. Frauke Urban's book doe
highly accessible way. The book is essential readi
who are interested in the Trias and who wan
nexus that shapes so much of current global poi.
— **Prof. Dr. Markus Lederer**, *Professor of International Relations and expert
in carbon governance, Technical University of Darmstadt, Germany*

"Whilst energy use remains tightly coupled to economic growth, how to square
the need for access to modern energy services with development challenges in
a carbon constrained world represents one of the most challenging conundrums
of our time - one of relevance across all of the UN's Sustainable Development
Goals. In this volume, Urban continues her track record of publishing accessible
introductions to core aspects of the field of climate change, energy and
development, once again offering a text that will be of value to both students and
teachers alike, as well as to informed policy makers and practitioners wishing to
extend their knowledge of this field. The inclusion of reflexive exercises at the
end of each chapter add to its pedagogic value, making it of particular value to
those designing new modules in this rapidly expanding field."
— **Prof. Dr. David Ockwell**, *Professor of Geography and co-convenor of the Climate and
Energy research domain in the ESRC STEPS Centre, University of Sussex, UK*

"This is a must read for all who are interested in energy, development and climate
change. An excellent reading resource addressing some of the world's most pressing
challenges, namely how to reduce energy poverty and achieve development, while
mitigating climate change."
— **Dr. Johan Nordensvärd**, *Senior Lecturer in Political Science and expert in the
social implications of energy and environment, Uppsala University and
University of Linköping, Sweden*

"This book comprehensively introduces the reader to the global transition that is
underway in energy production, consumption and access and the implications of
these changes for economic and human development. It recognises that progress in
tackling the twin problems of climate change and highly unequal access to energy –
energy profligacy alongside energy poverty – is proceeding too slowly. In a concise
but authoritative fashion the chapters examine the complex relationships between
the environmental, economic and political processes that shape this energy transition. Practical exercises at the end of each chapter encourage the reader to analyse
open access datasets and explore key policy documents and scientific assessments.
This excellent book is essential reading for students, academics and policymakers
wanting a multidisciplinary account of the ways in which energy and development
are interacting and what this means for the future."
— **Professor David Hulme**, *Executive Director, Global Development Institute
and FutureDAMS Research Programme, University of Manchester, UK*

ENERGY AND DEVELOPMENT

This book explores the complex relationship between energy and development and discusses the core issues and concepts surrounding this growing area of research and policy.

In the field of energy and development, the world faces two major challenges: (1) Providing energy access to the roughly one billion people worldwide who do not have access to electricity and the nearly three billion people worldwide who do not have access to clean cooking fuels; (2) achieving socioeconomic development while limiting global atmospheric temperature increases to 2 degrees Celsius to mitigate climate change. Taking stock of progress, Frauke Urban explores the key issues surrounding these goals and addresses the policy responses aimed at ending energy poverty and achieving sustainable development. She outlines various options for delivering energy access, analyses past and prospective energy transitions and examines the social, environmental, economic and technological implications of these possibilities. Taking a holistic and multi-disciplinary approach and containing useful teaching resources, *Energy and Development* provides a comprehensive overview of this complex field of study.

This book will be a great resource for postgraduate and undergraduate students, scholars, practitioners and policymakers working in the fields of energy studies, international development, environmental studies, industrial engineering, as well as social sciences that relate to energy and development.

Frauke Urban is Associate Professor in the management of sustainable energy systems at KTH Royal Institute of Technology in Stockholm, Sweden, and a Research Associate at SOAS, University of London, UK. She is an environmental scientist with more than 15 years' experience in research, teaching and publications related to energy, environment and development.

Rethinking Development

Rethinking Development offers accessible and thought-provoking overviews of contemporary topics in international development and aid. Providing original empirical and analytical insights, the books in this series push thinking in new directions by challenging current conceptualizations and developing new ones.

This is a dynamic and inspiring series for all those engaged with today's debates surrounding development issues, whether they be students, scholars, policy makers and practitioners internationally. These interdisciplinary books provide an invaluable resource for discussion in advanced undergraduate and postgraduate courses in development studies as well as in anthropology, economics, politics, geography, media studies and sociology.

Researching South–South Development Cooperation
The Politics of Knowledge Production
Elsje Fourie, Emma Mawdsley and Wiebe Nauta

Aid Power and Politics
Edited by Iliana Olivié and Aitor Pérez

Participatory Arts in International Development
Edited by Paul Cooke and Inés Soria-Donlan

Energy and Development
Frauke Urban

Power, Empowerment and Social Change
Edited by Rosemary McGee and Jethro Pettit

For more information about the series, please visit www.routledge.com

ENERGY AND DEVELOPMENT

Frauke Urban

Routledge
Taylor & Francis Group

LONDON AND NEW YORK

from Routledge

First published 2020
by Routledge
2 Park Square, Milton Park, Abingdon, Oxon OX14 4RN

and by Routledge
52 Vanderbilt Avenue, New York, NY 10017

Routledge is an imprint of the Taylor & Francis Group, an informa business

British Library Cataloguing-in-Publication Data
A catalogue record for this book is available from the British Library

Library of Congress Cataloging-in-Publication Data
Names: Urban, Frauke, 1980– author.
Title: Energy and development / Frauke Urban.
Description: Milton Park, Abingdon, Oxon ; New York, NY : Routledge,
   2020. | Series: Rethinking development | Includes bibliographical
   references and index. | Identifiers: LCCN 2019023590 (print) |
   LCCN 2019023591 (ebook) | ISBN 9781138485952 (hardback) |
   ISBN 9781138485969 (paperback) | ISBN 9781351047487 (ebook)
Subjects: LCSH: Energy policy. | Energy development. | Poor—Energy
   assistance.
Classification: LCC HD9502.A2 U73 2020 (print) | LCC HD9502.A2
   (ebook) | DDC 333.79—dc23
LC record available at https://lccn.loc.gov/2019023590
LC ebook record available at https://lccn.loc.gov/2019023591

ISBN: 978-1-138-48595-2 (hbk)
ISBN: 978-1-138-48596-9 (pbk)
ISBN: 978-1-351-04748-7 (ebk)

Typeset in Bembo
by Apex CoVantage, LLC

MIX
Paper from
responsible sources
FSC
www.fsc.org   FSC™ C013985

Printed in the United Kingdom
by Henry Ling Limited

To Jonathan and Emilie

— You are my sunshine —

CONTENTS

FOREWORD

This book has been in process for a long time. A different version of the material was first written by me for distance-learning teaching at SOAS, University in London, UK. Later, I re-wrote, updated and adapted the content into a textbook for teaching. This involved adding new material such as a new chapter on China and other emerging economies and their role for the global energy and climate systems, new information on latest low carbon technologies such as BECCS (bioenergy carbon capture and storage) and recent cost developments of solar PV and other renewable energy technologies. It involved thoroughly updating this fast-moving field by adding updates, especially after the introduction of the Paris Agreement and the IPCC's 1.5 degree report, as well as adding new thought-provoking, interactive exercises. I would like to thank SOAS University of London and especially Colin Poulton and Jennifer Yong, as well as other colleagues at the Centre for Development, Policy and Environment (CeDEP) and the Distance Learning Administration (DLA) for giving me the opportunity to work on this exciting project.

I would also like to thank KTH Royal Institute of Technology for their support regarding me writing this book, especially Cali Nuur. I am grateful for the time and space I was able to take to develop this project.

I am grateful to Matthew Shobbrook and Annabelle Harris from Taylor & Francis for their patience throughout this rather long process. Thank you to Aruna Rajendran and the copy editors.

Finally and most importantly, I would like to thank my family for supporting me. Emilie and Jonathan, you are both amazing and I am very grateful for having you. Johan, thank you for always supporting me and being there for me. Thank you to Eva and Reinhold for your lifelong support and encouragement.

1

ENERGY, POVERTY AND DEVELOPMENT

The challenges

Energy and development: why is it relevant?

Energy and development are inextricably linked. Energy is needed for basic human needs such as for cooking food, boiling water, space heating and cooling as well as lighting and other household activities. Energy is also necessary for income-generating activities like agriculture, industries and services. Energy is crucial for enabling modern transportation, Information and communications technology (ICT), healthcare and sanitation. Yet, about 1 billion people worldwide, mainly in rural areas, do not have access to electricity, and about 2.7 billion people rely on traditional biomass – such as fuelwood, charcoal, crop residues and dung – for basic needs such as cooking and heating (IEA, 2019a).

While challenges such as energy poverty continue to exist, a lot of progress has been made over the last few decades, and large-scale changes have occurred:

First, providing modern energy for human development is today a global, regional and national priority. The Sustainable Development Goals (SDG), particularly the energy goal SDG 7, and the United Nations' Sustainable Energy for All (SE4All) initiative are aiming to provide universal modern energy access to everyone worldwide by 2030. These global targets have been translated into national targets and remain a priority for many countries.

Second, electricity generated from renewable energy such as solar photovoltaics (PV), wind energy, hydropower and modern biomass are today economically competitive compared with electricity generated from fossil fuels. The price drop for solar PV has been particularly striking in the last few years. Renewable energy has established itself as an important part of the global energy mix over the last couple of decades, and more and more countries are diversifying their energy mix by relying on their domestic resources of renewable energy.

Third, the development of emerging economies like China, India, Brazil and South
Africa has not only changed the world geopolitically and in terms of economic
power, but also in terms of environmental impact and environmental leadership. This is very clear in the field of energy, climate and development. Most
striking is the development of China, which emerged into an economic, political, demographic and also environmental global power. Since 2007, China has
been the world's largest emitter of carbon dioxide (CO_2) and the world's largest
energy consumer. Today China accounts for about 20% of global energy consumption and about 30% of global CO_2 emissions and has a high dependency
on polluting coal – making up nearly 70% of the country's electricity generation (IEA, 2019a). On the other hand, China is also the world's largest investor
and installer of renewable energy including hydropower, wind energy and solar
PV. Its policymaking is actively driving forward climate action. China's domestic energy and climate policy therefore matters globally, and the country has
also become a major player in the international climate negotiations.

Fourth, after many years of standstill, the international community managed to
negotiate a universal climate agreement for all countries, the Paris Agreement.
It aims to avoid dangerous climate change by limiting atmospheric temperature
increases to below 2 degrees Celsius. Many challenges remain as currently the
policy ambitions are not in line with the 2 degrees goal and global emissions
are still increasing rather than decreasing. Energy plays an important role in
climate change mitigation, and the next few years will be crucial for advancing
low carbon energy transitions in all sectors to mitigate global climate change.

The purpose of this book

Energy and development are therefore at the heart of wider international, regional,
national and local policy agendas that aim to deliver human development, while at
the same time living within the planetary boundaries. This book serves as an introduction to and an overview of the field of energy and development. It addresses the
issues raised above and many more. Many issues are highlighted both conceptually
and theoretically, as well as in applied terms that are relevant for real-life challenges
faced by governments, businesses, donors, NGOs, institutions, the scientific community and the public. The textbook is relevant for master's students, advanced
undergraduate students, postgraduate students, as well as academics, policymakers
and practitioners who are interested in energy and development issues. The book is
written from an interdisciplinary perspective, combining relevant knowledge from
development studies, environmental studies, engineering, economics and social
sciences.

After each chapter there are thought-provoking, interactive exercises. These
exercises are not merely testing or repeating the knowledge that has been read,
but they rather encourage the development of the reader's own analytical skills,
critical thoughts and further understanding by applying the chapter's content to
real-life cases. For example, the exercises will ask for applying specific concepts or

theories to real cases by analysing data from the World Bank or the International Energy Agency; or, they will ask for exploring a specific case for which a presentation should be made or a speech should be prepared, as if the reader were a specific actor (e.g. a policymaker urging their parliament to take action, or an organisation bidding for donor funding). The aim is to build up the reader's understanding of the subject area by first reading about it and then actively applying it.

The link between energy and development

What is energy?

Energy is a physical term that describes the capacity of a physical system to perform work. Energy exists in several forms such as thermal energy (heat), radiant energy (light), mechanical energy (kinetic), electric energy, chemical energy, nuclear energy and gravitational energy. There is also a difference between *potential energy* that is being stored, such as the water in the reservoir of a hydropower dam, and *kinetic* or *working energy*, such as the energy produced when the water is released and the turbines are operating (Cutnell & Johnson, 2012).

Energy is also a physical unit. It is usually measured in joules (J). In physics, the law of conservation of energy suggests that within a closed system the total energy remains constant and cannot change. This means that energy is conserved. Energy cannot be created or destroyed; however, it can change its form (Cutnell & Johnson, 2012). An example is thermal energy – such as from a thermal coal-fired power station – that is being converted into electric energy; or kinetic energy – such as from the water of a hydropower dam that produces electric energy (Goldemberg & Lucon, 2009).

Energy carriers and energy sources can be differentiated. An *energy carrier* is a substance or system that contains potential energy that can be released and used as actual energy, in the form of mechanical work, for heating or for operating chemical and physical processes. For example, energy carriers include batteries, coal, dammed water, electricity, hydrogen, natural gas, petrol and wood. Energy carriers do not produce energy; however, they 'carry' the energy until it is released.

Energy sources can be divided in renewable and non-renewable energy sources. The term refers to the resources that are being used for the energy, for example coal or wind. *Renewable energy* resources are abundantly available and can be renewed over time, such as energy from wind, the sun (solar), water (hydropower) and biomass. Non-renewable energy sources come from resources that are finite and can be depleted, such as *fossil fuel energy* resources like coal, oil and natural gas, but also nuclear energy, such as uranium (Goldemberg & Lucon, 2009).

We can differentiate between primary and secondary energy. *Primary energy* has not been subject to any conversion or processing and contains raw fuels, such as crude oil or solar energy. *Secondary energy* has been subject to conversion or processing, such as from crude oil to petrol for powering vehicles. Another example is the conversion of solar power to electricity (Goldemberg & Lucon, 2009).

What are W, kW, MW, GW, TW and PW and how do they relate to energy?

Answer
These are units of power that measure the rate of energy conversion in the International System of Units (SI). The basic unit is expressed in watts (W), named after the inventor James Watt (1736–1819). One watt equals one joule per second.
> *watt = 1 W*
> *kilowatt = 1 kW = 10^3 W*
> *megawatt = 1 MW = 10^6 W*
> *gigawatt = 1 GW = 10^9 W*
> *terawatt = 1 TW = 10^{12} W*
> *petawatt = 1 PW = 10^{15} W*

The next section will discuss the term development, followed by a section on the links between energy and development.

What is development?

There are various definitions for *development* and despite its universal use, there is no universally agreed definition. Some scholars such as Robert Chambers simply define development as '"good" change' (Chambers, 1995: 174); others associate it with progress and/or modernisation, or economic growth. Even others make a distinction between formal development, such as development aid, and development as a deeper process of change, such as capitalism (Urban et al., 2011). Hart (2001: 650) distinguishes between 'big D' and 'little d' development whereby

> 'big D' development [is] defined as a post-second world war project of intervention in the 'third world' that emerged in the context of decolonization and the cold war, and 'little d' development or the development of capitalism as a geographically uneven, profoundly contradictory set of historical processes.
> *(cited by Urban et al., 2011: 6–7; Urban & Nordensvärd, 2013: 10)*

There are various approaches to Western development thinking, including rights-based approaches, which focus on human rights and/or increasing the voice of marginalised groups (Mohan & Holland, 2001; Hickey & Mohan, 2005; Urban et al., 2011). There are also human development approaches, which incorporate broader development objectives than economic ones and aim to expand human choices and strengthen human capabilities related to education, health, income and gender (Jolly, 2003; Urban et al., 2011). There are also approaches that are based on concerns for the poorest 'bottom billion' (Collier, 2007; Urban et al., 2011). There are further approaches that come from different disciplines such as anthropology, economics and political science, and different perspectives such as gender, globalisation and the environment. In other parts of the world, such as China, different streams of non-Western development thinking prevail which are more related to these countries' own experiences, culture and philosophy of development (Urban et al., 2011; Urban & Nordensvärd, 2013).

While humans have been concerned about economic development and social transformation for centuries, the concept of international development and *development studies* as a discipline is reported to have emerged in the late 1940s, 1950s and early 1960s. Development studies began as a post-Second World War project in support of poorer 'developing countries'. 'Development' was driven by so-called developed Western/Northern countries. 'Development' has often been accused of paternalism and trusteeship (Cowen & Shenton, 1996; Urban et al., 2011). Back in the 1950s, development policy was dominated by the goal of achieving modernity, by an optimist worldview, by expecting the state to play an active, positive role and by focussing on national development (Humphrey, 2007; Urban et al., 2011; Urban & Nordensvärd, 2013).

While development studies started with optimism after the Second World War, the concept of development, and development studies as a discipline, has endured criticism for many years (Urban et al., 2011). This is linked to ongoing problems such as widespread poverty in many parts of the world; global neoliberalism, which sees states as part of the problem rather than part of the solution (Humphrey, 2007); as well as the occurrence of various transboundary phenomena. Challenges like the global financial crisis, terrorism and large-scale environmental problems, such as climate change and natural resource depletion, are seen to require international and multilateral solutions (Urban et al., 2011). One other major shift in development policy is due to the rise of emerging economies like China, India, Brazil, South Africa and states of the Middle East (Urban et al., 2011). Their rise questions dominant 'Western' approaches to development (Humphrey, 2007). Unfortunately, the optimism of earlier decades has been replaced by some pessimism, including development being declared dead in the 1990s by both the political right and the political left (Hart, 2001; Urban et al., 2011). Fifteen years later, Rist argued that development as practised and imposed by the West was 'toxic' (Rist, 2007; Urban et al., 2011; Urban & Nordensvärd, 2013).

The notion of 'reimagining development' therefore prevails within the discipline of development studies, with new thinking of what development policy and practice means today, who is driving it and for whom, particularly in view of the Sustainable Development Goals (SDGs) to be achieved by 2030. In the field of energy and development, the most important initiatives are SDG 7 on sustainable energy and the UN energy access initiative SE4All that provides positive and forward-looking pathways to achieving access to sustainable energy for everyone worldwide to enable sustainable development (SE4All, 2014).

Please note: The terms 'developing' country and 'developed' country are used in line with international practice. Nevertheless, the author acknowledges the following: First, there is a wide range of so-called developing countries, ranging from the least developed countries to low-income countries, lower-middle-income countries, upper-middle-income countries and emerging economies. Some of these terms overlap. For example, China is an emerging economy and an upper-middle-income country, whereas Haiti is a least developed country and a low-income country. However, these categories are changing from year to year, meaning that one year a country can be in the low-income group and the next year it can be classified by the World Bank as a middle-income country. Second, the categorisation into 'industrialised countries' is not helpful either as some emerging economies, such as China, are increasingly industrialised and are no longer predominantly agrarian-based economies. Third, there is a geographical confusion concerning the global 'North' and the global 'South'. The global North is often referred to as developed, industrialised countries including mainly North America (US and Canada), Europe (the EU), Australia (Australia and New Zealand) and Japan. Obviously, Australia, for example, is situated in the southern hemisphere; therefore, this classification is false. The global South includes all developing countries; however, geographically many of these countries are not located in the southern hemisphere, such as some countries located in Asia, Latin America and Northern Africa. Fourth, these classifications often conceal information about income distribution. A country such as Angola may currently be classified as a lower-middle-income country (and previously it was classified as an upper-middle-income country). Yet, this does not imply that most of its citizens would be classified as living in a middle-income country. The reality is that stark differences exist between the poor and the rich in a mineral-rich country such as Angola, thereby creating an average middle-income classification that conceals the realities for many of its citizens. Fifth, many of these classifications, such as developing versus developed, industrialised versus agrarian, South versus North, East versus West are outdated and stem from a time when globalisation did not exist to the same extent as today and the world was much easier to classify, both in terms of income and in terms of ideologies (e.g. East versus West). These classifications for grouping countries are flawed when considering the multipolar world we live in today. In the absence of a valid alternative, the author has chosen to refer to developed/industrialised and developing countries as well as low income, middle income, high income, emerging economy and specific country names as appropriate.

Energy and development

Energy is used for virtually everything. Energy is required for basic human needs: for cooking, lighting, boiling water, heating and cooling, and for other household activities. Energy is also required to sustain and expand economic processes such as agriculture, electricity production, industries, services and transport. Energy is also needed for healthcare, for telecommunications and to provide clean water and sanitation.

In line with this statement, it is commonly suggested that *access to energy* is closely linked with development and economic growth (e.g. UNDP & WHO, 2009; IEA,

2010; SE4All, 2014). This has been acknowledged in the UN Sustainable Energy for All Initiative (SE4All), which aims to provide access to modern energy services to everyone worldwide by 2030 (as well as double the rate of improvement of energy efficiency and double the share of renewable energy in the energy mix) (SE4All, 2014). Access to modern energy services is important for reducing the time, burden and danger of fuelwood collection, to power industries and services that generate economic growth and to increase energy security.

When we talk about *modern energy services*, we refer to options other than *traditional biomass* (such as fuelwood, dung, agricultural residues and charcoal) and refer instead to electricity and other modern energy options such as biogas. This will be discussed in more detail in the following sections and units.

Access to modern energy services is therefore crucial for development. There are, however, various ways of defining energy access. Practical Action defines 'total energy access' as having access to energy for lighting, cooking and water heating, space heating and cooling, information and communications, and energy for earning a living (Practical Action, 2010). Partial energy access is defined as having access to energy for some of these activities, for example energy for cooking but not for lighting. There is therefore a differentiation between access to electricity, access to household fuels (such as for cooking) and access to mechanical power (such as a treadle pump for pumping water).

Access to modern energy services is often equated with access to electricity. The definition of *electrification* differs between various institutions and countries. In principle, a household or village should only be classified as electrified once everyone in the household or village has access to reliable electricity. This is, however, not the case. For example, the Indian government uses the definition that a village or neighbourhood is electrified when at least 10% of the households have access to electricity. Other interpretations by different actors are that a village or neighbourhood is electrified if electricity is being used for any purposes. This may not necessarily mean that people have access to electricity at home; it may mean, however, that at least one household or building (e.g. a school, a clinic) has access to electricity. Statistics on electrification rates therefore have to be viewed with caution as they might overestimate the actual number of people who have access to electricity. The electrification rates also do not give information about the reliability of the supply, as power cuts (blackouts or load shedding) are common in many countries around the world. Bear this in mind when reading the following paragraph and looking at Figure 1.1.

There is a *link between income levels and energy access*. A correlation has been found between rising income levels, both at household level and at national level, and rising energy access. This is valid for electricity access and access to modern fuels (World Bank, 2014). Consequently, countries which have higher incomes tend to have higher electricity access rates (Urban & Nordensvärd, 2013). While a clear trend can be seen, there are always exceptions. An exception, for example, is Vietnam, which has still relatively low income levels in comparison to other Asian countries, but has an electrification rate of about 100%. This is due to consolidated government efforts over many years for national electrification. The addition of a lot

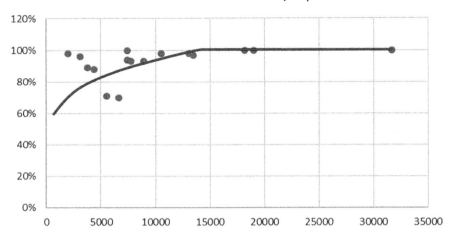

FIGURE 1.1 Electrification rates in Asian developing countries versus GDP/capita (purchasing power parity PPP) based on data from 2019

Source: Data on electrification and GDP/capita PPP from the IEA (2019a) and World Bank (2019)

Note: The *x* axis shows GNI/capita (PPP) in US$; the *y* axis shows electrification rates in %. Please note that progress in electrification is rapidly being made so that the actual percentages may increase every year, as long as socio-economic development continues.

of new hydro-electric capacity in countries such as Cambodia, Lao PDR, Myanmar and Bhutan has also significantly increased electrification rates. Figure 1.1 shows the national electrification rates in countries of developing Asia. The countries are displayed in the graph according to their gross domestic product (GDP) per capita purchasing power parity (PPP) and their electrification rates in percentages. It is evident that countries that have higher per capita incomes, such as Malaysia, China and Thailand, have higher electrification rates, whereas countries with lower per capita incomes, such as Cambodia, Bangladesh, Pakistan and Myanmar, have lower electrification rates. Yet, rapid electrification has occurred across developing Asia in recent years, which is a very positive step forward.

Energy and poverty

Despite the importance of energy access, about 1 billion people worldwide do not have access to electricity and 2.7 billion people rely on traditional biomass – such as fuelwood and dung – for basic needs such as cooking and heating (IEA, 2019a). This means that about 20% of the global population does not have access to electricity, although in many developing countries the figure can be as high as 80% or 90%. About 85% of these people without access to electricity live in rural areas and 95% of them live in sub-Saharan Africa and developing Asia. About 40% of the global population relies on traditional biomass, although in many developing countries the figure is much higher (World Bank, 2014).

Energy poverty is defined as the lack of access to electricity and a reliance on the traditional use of biomass for cooking (IEA, 2010; Practical Action, 2010).

There are two so-called hotspots of energy poverty: one in sub-Saharan Africa and one in developing Asia. In sub-Saharan Africa, almost 70% of the population does not have access to electricity and 80% rely on traditional biomass. The entire population of sub-Saharan Africa (excluding South Africa) – about 800 million people – used as much electricity as the 20 million people of New York, USA in 2010. The share of access to electricity and modern energy is higher in developing Asia; however, a particular energy poverty hotspot is located in India, where more than 150 million people do not have access to electricity, despite rapidly increasing electrification rates (IEA, 2019b). Energy poverty is therefore widespread and poses a global development challenge.

While the technical term used in developing countries is 'energy poverty', the technical term often used to describe a similar situation in developed countries is 'fuel poverty'. An individual or a household lives in fuel poverty when they cannot afford to pay to keep adequately warm in their home given their low income. The term is mainly used in the UK, Ireland and New Zealand, although similar discussions exist across Europe and the US. In the UK, an individual or a household is classified as living in fuel poverty when they spend 10% or more of their income on the costs of heating (UK Government, 2000). Fuel poverty can have several causes: low income, high fuel prices, poor energy efficiency of homes or under-occupancy of homes. It is often suggested that those affected the most are the elderly, as they often tend to live on low pensions in homes that are not of the latest energy efficiency standard.

Figure 1.2 shows the unequal distribution of people without access to electricity. The white fields indicate countries that have 100% access to electricity. The coloured fields show how many million people do not have access to electricity, with grey representing those countries where large numbers of people still have no access to electricity, such as in India and parts of sub-Saharan Africa.

Figure 1.3 shows the unequal distribution of people without access to clean cooking fuels. The white fields indicate countries that have 100% access to clean cooking fuels. The coloured fields show how many million people do not have access to clean cooking fuels, with dark grey representing those countries where large numbers of people still have no access to clean cooking fuels, such as in India, China and parts of sub-Saharan Africa.

Dealing with energy poverty

Measuring energy poverty

This section deals with *energy poverty* and how to measure it. Before delving into the theories and practicalities of how to measure energy poverty, we will first look at two real-life examples of what energy poverty means to poor people.

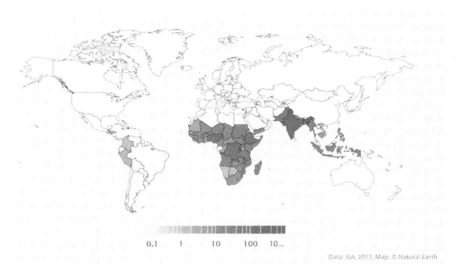

FIGURE 1.2   Population without access to electricity, measured in millions

Data source: IEA Energy Access Database, 2019b. www.iea.org/energyaccess/database/ All rights reserved. This map is without prejudice to the status of or sovereignty over any territory, to the delimitation of international frontiers and boundaries and to the name of any territory, city or area.

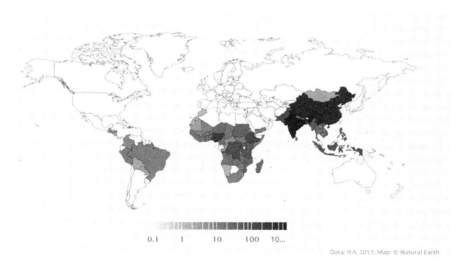

FIGURE 1.3   Population without access to clean cooking fuels, measured in millions

Data source: IEA Energy Access Database, 2019b. www.iea.org/energyaccess/database/ All rights reserved. This map is without prejudice to the status of or sovereignty over any territory, to the delimitation of international frontiers and boundaries and to the name of any territory, city or area.

Rosa from Kenya says:

> For me getting energy for cooking and lighting is a daily worry. It's so hard to find firewood that I cook for my family only once a day, in the evening. The fire provides the light for cooking and eating a meal with my children. After eating is bedtime.
>
> *Practical Action (2010: v)*

Maya from Nepal reports:

> We are totally dependent on firewood for cooking energy as we don't have other alternatives. Managing firewood is a very tedious job for us. We have to walk about 7 hours to collect a Bhari (about 30 kg) of firewood. The track to the forest is very difficult and unsafe.
>
> *Practical Action (2010: 13)*

Keeping in mind how difficult it is to live with energy poverty, we will discuss how to measure it and then how to overcome energy poverty by providing energy access.

There are various ways of measuring energy poverty. Two very simple and rather crude measurements are *electrification rates* and the *rates of people cooking with traditional biofuels* such as fuelwood, charcoal, dung and agricultural residues.

Another useful indicator is the *Multidimensional Poverty Index (MPI)*. The MPI is not a pure energy indicator, but it measures development levels from a holistic perspective, taking into account energy poverty indicators as some of several development indicators. The MPI measures the lack of access to electricity and the prevalence of traditional biofuels for cooking such as fuelwood, charcoal or dung (OPHI, 2010).

A further index for measuring energy poverty is the *(Total) Energy Access Index* (Practical Action, 2010). This index specifies the minimum energy standard individuals or households should have for specific energy services, such as lighting, cooking and water heating, space heating, cooling, information and communications, and earning a living.

The Energy Access Index then ranks access to energy services on levels from 1 to 5 for access to household fuels, electricity and mechanical power. Depending on the quality of the supply this can, for example, range from no access to electricity (level 1) to reliable 240 V alternating current (AC) connection for all uses (level 5) (Practical Action, 2010).

It is important to understand how energy poverty is measured to be able to understand the underlying issues that cause energy poverty and the impacts of energy poverty, such as health problems from indoor air pollution, the burden of fuelwood collection, limited opportunities to earn a living, limited access to communications and services, etc.

Policy and practice for overcoming energy poverty

As about 1 billion people worldwide do not have access to electricity and 2.7 billion people rely on traditional biomass for basic needs such as cooking and heating (IEA, 2019a), there is an urgent need to overcome energy poverty. As mentioned in earlier sections, energy poverty is linked to low national per capita incomes, to low development levels and to a wide range of social, economic and environmental challenges. Overcoming global energy poverty is therefore one of the biggest challenges of our times. This is recognised by the fact that energy is relevant for all of the Sustainable Development Goals, and most of them cannot be achieved without energy access. The SDGs are 17 development goals that were launched by the United Nations (UN) in 2015 and have the aim to fundamentally improve the world in a sustainable way by 2030, such as by eradicating poverty, eliminating hunger, improving health and well-being, improving quality education, enabling gender equality, providing clean water and sanitation, providing affordable and clean energy to everyone worldwide, tackling climate change and achieving several other important goals. Fortunately, the understanding of energy poverty has improved in recent decades, and its importance has gained some global attention in recent years, hence the SDGs present an opportunity to include energy targets, namely target 7 on affordable and clean energy. More on the SDGs and energy issues will be discussed in the forthcoming chapters.

Efforts to reduce energy poverty and increase energy access have been ongoing for several decades; however, they have had limited success. While many country- and local-level initiatives aimed (and still aim) at alleviating energy poverty and increasing energy access, there have been no far-reaching international energy access targets in the past. This has changed with the United Nation's targets for universal energy access. The UN made 2012 the International Year of Sustainable Energy for All and has set a target for providing universal modern energy access by 2030, which is in line with SDG 7. This initiative is called Sustainable Energy for All (SE4All). This target aims to provide access to electricity and to clean cooking facilities to everyone in every country worldwide. This goal is linked to renewable energy provision, as the UN estimates that about two thirds of the rural population in developing countries will get access to electricity through *decentralised renewable energy*, such as from wind, solar and small hydro. This will be delivered by renewable-energy-powered *mini-grids* and *off-grid* solutions. Decentralised renewable energy, such as biogas, also plays a key role in providing the rural poor with access to clean cooking facilities (IEA, 2010). The UN General Assembly also declared 2014–2024 the Decade of Sustainable Energy for All.

One of the key questions is how the universal energy access targets will be financed. It is suggested that a portion of the costs will have to be borne by national governments and authorities, another portion may be funded by the international community and NGOs and yet another portion by the private sector as a means of investing in energy infrastructure and technology. Nevertheless, some of the costs will be borne by consumers, who in this case are predominantly poor and have limited ability to pay (IEA, 2011).

TABLE 1.1 Targets for the UN's universal modern energy access to be achieved by 2030

Goal	By 2030	
	Rural area	**Urban area**
Providing access to electricity	100% of population has access, of which 30% is connected to the grid and 70% either mini-grid (75%) or off-grid (25%)	100% access to grid
Providing access to clean cooking facilities	100% access to liquid petroleum gas (LPG) (30%), biogas systems (15%) or advanced biomass cook stoves (55%)	100% access to LPG stoves

Source: IEA (2010: 116)

Table 1.1 indicates the targets for the SE4All initiative and how energy access should be provided in the urban and rural areas.

After examining international policy and practice for overcoming energy poverty, we will briefly examine examples from two national governments.

We will first look at India. With more than 150 million people without access to electricity, India hosts the world's largest population deprived of electricity. The large majority of this population lives in rural India (IEA, 2019b). Yet, significant progress has been made within the last decade, hence the number of people without access to electricity was roughly halved between 2007 and 2017 (IEA, 2019a). The World Bank reported that nearly 93% of the population in India had access to electricity in 2017 (IEA, 2019a). However, the Indian government claims that in April 2018 all villages were electrified and by 2020 everyone should have access to electricity. Yet, this announcement has to be viewed with caution, as this is unlikely to mean that every person in the village has access to electricity. Instead, the Indian village electrification scheme defines villages as electrified when 10% of its homes and all public buildings have access to electricity (BBC, 2018). There are several bottlenecks with regards to electrification as India has for decades suffered from an underfinanced power sector, poor infrastructure, customers who are too poor to pay, electricity theft and problematic restructuring of mostly state-owned utility firms (IEA, 2002, 2012).

After looking at India, the country hosting the world's largest population without electricity access, we will look at China. China has achieved an electrification rate of almost 100% in recent years due to a decade-long history of providing rural electricity access through small-scale hydropower and recent large-scale government-funded electrification programmes. The first large-scale rural electrification initiatives in the 1950s to 1970s were based on small-scale hydropower and aimed at increasing agricultural productivity by means of improved irrigation, driven by a centrally planned state. Over time, the policy on rural electrification became driven by local governments that made considerable investments, thereby helping to roll out rural electrification efforts all over the country as a means of providing access to electricity for millions of people. This happened at a time when China shifted from being mainly agrarian based to becoming increasingly industrialised in the 1980s and 1990s. Since 2000, the aim has been to invest in upgrading rural grids and extending electricity access to remote areas of China (Jiahua et al., 2006). In recent years, China launched the China Township Electrification Programme to provide electricity from renewable energy to 1000 townships, which was one of the largest of such programmes worldwide. This was followed by the China Village Electrification Programme, which aimed to bring electricity from renewable energy to 3.5 million households and to achieve complete electrification by 2015. Renewable energy, particularly small hydropower, has therefore always played a prominent role for reducing energy poverty in China. At the same time it has to be acknowledged that China sets itself apart from many other countries due to its formerly centrally planned state and what Chinese may today call a 'market economy with Chinese characteristics' that has abundant investments, well-operating state-owned enterprises and access to modern power technology.

The two examples above highlight some historic and current developments in the policy and practice of overcoming energy poverty.

Energy and environmental problems

Energy and climate change

About 80% of the global primary energy supply comes from fossil fuels, primarily oil and coal (IEA, 2019). Fossil energy resources are limited and *fossil energy use* is associated with a number of negative environmental effects, most importantly *global climate change*, but also natural resource depletion and air pollution. The next section discusses how energy use and climate change are linked.

The IPCC estimates that about 70% of all *greenhouse gas (GHG)* emissions worldwide come from energy-related activities. This is mainly from fossil fuel combustion for heat supply, electricity generation and transport, and includes carbon dioxide, methane and some traces of nitrous oxide (IPCC, 2007). It is well documented that these emissions contribute to global climate change. Energy use has potentially significant climate impacts, which are assumed to exceed the impacts from other sources like land use and other industrial activities. It is therefore considered crucial to promote GHG emission reduction technologies such as renewable energy to achieve carbon neutrality in the long term.

The IPCC states that the

> atmospheric concentrations of carbon dioxide, methane, and nitrous oxide have increased to levels unprecedented in at least the last 800 000 years. CO_2 concentrations have increased by 40% since pre-industrial times, primarily from fossil fuel emissions and secondarily from net land use change emissions.
> *(IPCC, 2013: 7)*

Energy use from fossil fuels therefore contributes directly to GHG emissions and thereby contributes to rising temperatures, melting glaciers and ice sheets, sea-level rise, extreme weather events and other observed effects of global climate change.

The IPCC has an extremely high confidence level of 95% probability that global climate change is anthropogenic, caused by excessive GHG emissions (IPCC, 2013, 2014a, 2014b). At the global scale, the atmospheric concentration of CO_2 has increased from a pre-industrial value of approximately 280 parts per million (ppm) in about 1860 to around 380 ppm in 2005 (IPCC, 2007), 396 ppm in 2007 (Richardson et al., 2009), just over 400 ppm in summer 2013 and about 412 ppm in June 2019 (NASA, 2019). See Figure 1.4 for details.

According to the IPCC (2018), the global mean surface temperature has risen by 0.8–1.2 degrees C between 1880 and 2017 (IPCC, 2018). This increase has been particularly significant over the last 50 years. From a global perspective, the IPCC (2013, 2014a, 2014b, 2018) reports that they found high increases in heavy precipitation events, while droughts have become more frequent since the 1970s, especially in the (sub)tropics. They also document changes in the large-scale atmospheric circulation and increases in tropical cyclone activity since the 1970s (IPCC, 2013, 2014a, 2014b). The IPCC's latest Fifth Assessment Report highlights the observed and partly irreversible changes to the earth's ecosystems, particularly the changes to the oceans, that absorb a large part of the CO_2 and thereby become acidified, and the cryosphere (IPCC, 2013, 2014a, 2014b; Urban, 2014). More recent scientific reports highlight the rapid and dramatic rate at which glaciers, ice sheets and ice

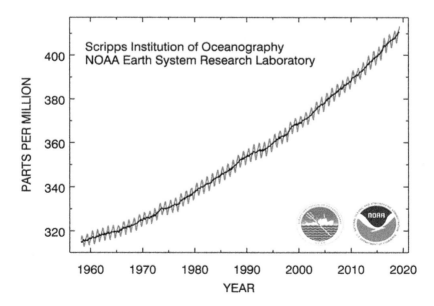

FIGURE 1.4 Atmospheric CO_2 concentrations in parts per million (ppm) at Mauna Loa Observatory between 1960 and 2018, reported in May 2019

Source: Tans and Keeling, NOAA (2019)

caps are melting, particularly in Antarctica and the Artic, the increased frequency and severity of extreme weather events across the globe such as storms, droughts, heat waves and wild fires, increased sea-level rise, as well as more wide-spread coral bleaching as the oceans are warming more quickly.

Today, the majority of climate scientists agree that 'the possibility of staying below the 2 degree Celsius threshold by 2100 between "acceptable" and *"dangerous" climate change* becomes less likely as no serious global action on climate change is taken' (IPCC, 2018; Urban, 2014: 4 also citing Anderson, 2009; Richardson et al., 2009). Climate scientists estimate that for a 50% chance of achieving the 2 degree target, let alone the 1.5 degree target advocated for by the IPCC (2018), a global atmospheric CO_2 equivalent concentration of 400 to 450 ppm should not be exceeded (Richardson et al., 2009; Pye et al., 2010). This would require an immediate reduction of global GHG emissions of about 60%–80% by 2100 (Richardson et al., 2009). Nevertheless, the 400 ppm target is reported to have been exceeded, with atmospheric CO_2 emissions by NASA being recorded as about 412 ppm in June 2019 (NASA, 2019). Still, emissions are rising. There is therefore a need to reduce emissions rapidly and significantly to avoid dangerous climate change (Urban, 2014).

The current scientific consensus after the UNFCCC Paris Agreement is that the current Nationally Determined Contributions (NDCs) of the signatory nations that outline their mitigation actions are not sufficient to keep a global temperature increase below 2 degree Celsius, let alone the 1.5 degree target that has been suggested by the Paris Agreement and the most recent IPCC report (IPCC, 2018). Rather, leading scientists estimate that the average warming will be between 2.6 and 3.1 degrees Celsius by 2100 or even higher (Rogelj et al., 2016). Knutti et al. (2016) argue that the target of 2 degrees (or especially the 1.5 degrees target) is unrealistic and not in line with real-world developments as the required emission reductions are still not happening.

Energy from non-fossil fuels, such as renewable energy and low carbon energy, is therefore crucial for mitigating the GHG emissions that lead to climate change. Other strategies such as reducing energy use and increasing energy efficiency are also required, as well as safeguarding carbon sinks such as forests. In the long term, carbon neutrality needs to be the ultimate goal, which means restructuring of all industrial sectors.

Energy and other environmental problems

Natural resource depletion

Another key environmental impact of energy use is *natural resource depletion*. Energy resources are a form of natural resource. One differentiates between fossil and non-fossil resources. *Fossil resources* have been formed over millions of years from the organic remains of prehistoric animals and plants. They have a high carbon content and include coal, oil and natural gas. These fossil fuels are non-renewable energy sources, as their reserves are being depleted much faster than new ones are being formed. For example, it took millions of years to form an oil field, but it can take only several years of exploitation to deplete it. Mining and extraction are therefore major causes for depleting fossil fuel resources (Goldemberg & Lucon, 2009). It has been estimated that two thirds of fossil fuel resources need to stay in the ground to avoid dangerous climate change.

Non-fossil resources include renewable non-finite, non-depletable energy such as wind energy, solar energy and hydropower that are abundantly available from the wind, sun and water. Renewable energy comes from the earth's elements that are always available, such as sun and wind.

Biomass-based energy is another of energy source that is being classified as renewable. Nevertheless, there is a division between traditional biomass such as fuelwood, modern biomass such as pellets, wood chips and wood briquettes, biogas such as for cooking, as well as biofuels that are divided in first-generation, second-generation and third-generation groups. First-generation or conventional biofuels are usually based on edible biomass-based starch, sugar or vegetable oil. This means these fuels are usually based on food products such as corn, wheat or other cereals, cassava, sugar beet and sugar cane, which are used for making bioethanol. Soy, jatropha and palm oil are used for making biodiesel. Second-generation biofuels are usually made from feedstock and waste (e.g. municipal waste). Third-generation biofuels or advanced/unconventional biofuels usually do not depend on food products or feedstock, but can be derived from algae, cellulose and other forms of plant biomass, which makes it harder to extract fuel (Goldemberg & Lucon, 2009).

Nuclear energy also falls into the category of non-fossil resources. While nuclear energy is a non-fossil resource, it is finite and depletable as it relies on uranium resources. There is heated debate on whether uranium will become a rare and near-depleted resource, similar to oil, any time soon or whether uranium resources will remain abundant for hundreds or even thousands of years to come.

At the same time, the *harvesting and use of traditional biomass* such as fuelwood can contribute to the depletion of *natural forest resources*. While many people rely on the collection of fallen down branches from trees, others depend on felling trees. The production of charcoal also involves the felling of trees. This can lead to a degradation and/or decrease of woodlands and forests that can eventually lead to larger-scale deforestation, erosion and desertification. This practice can also lead to a decrease in biodiversity and negative impacts on water and food security. In areas where forests offer some protection again natural disasters, felling trees for access to fuelwood can have devastating impacts. For example, as mangrove forests are cleared for fuelwood, this reduces the natural protection from floods, tsunamis and sea-level rise, thereby leaving people, economies and ecosystems even more vulnerable to natural disasters and climatic impacts than before.

Another key issue around energy use and natural resource depletion is linked to peak oil. *Peak oil* is a concept that describes first an increase in oil production up to a peak and afterwards a decline in oil production. This is based on an observed rise of oil production, a peak and then a fall in the production rate of oil fields over time as the oil resources are depleted. The theory is that this phenomenon of peaking oil production is not only limited to oil fields, but also applies globally as all oil resources could be depleted at some point. This is due to rapid oil extraction from finite natural resources that can be depleted and that needed millions of years to be built. Peak oil is therefore the point of maximum oil production. There is lively debate among scientists and the oil lobby over whether peak oil has already happened or whether it still lies ahead. Some experts argue that we are either approaching or already beyond peak oil as the rate of depletion is rapid, oil prices have increased to formerly unseen levels in recent years, and few new conventional oil resources are being found. Yet, the extraction of so-called unconventional oil resources has expanded rapidly, such as from shale gas, shale oil and tar sands. The extraction of unconventional oil resources involves environmentally destructive methods such as fracking (Leggett, 2013). The UK Energy Research Centre published a report that reviewed over 500 studies on peak oil and global supply forecasts and concludes that a peak in oil production is likely to happen before 2030, if not earlier (UK ERC, 2009).

Air pollution

Energy use is not only likely to contribute to global climate change, but also gives rise to other negative atmospheric impacts such as *air pollution* (Goldemberg & Lucon, 2009). This includes indoor air pollution and outdoor air pollution.

Indoor air pollution is caused by the combustion of traditional solid fuels such as fuelwood and charcoal. The combustion of these solid fuels causes indoor air pollution through the release of smoke, soot and small particles that are linked to negative impacts on health. This is associated with pneumonia, chronic respiratory disease, lung cancer and adverse pregnancy outcomes due to exposure to indoor air pollution (WHO, 2000, 2005, 2006). The World Health Organization (WHO) reports that about 5% of all deaths in least developed countries could be due to traditional solid fuel use (WHO, 2000). According to WHO (2019), about 3.8 million people – mostly women and children who spend much of their time close to the hearth – are likely to die every year because of exposure to indoor air pollution from traditional biofuels. This is more people than die each year from malaria or tuberculosis. Introducing modern renewable and low carbon energy sources as a replacement for traditional biofuels is likely to increase the health of the population in developing countries.

Outdoor air pollution is another observed phenomenon that is linked to the combustion of fossil fuels from energy generation, transport and industry. Outdoor air pollution can create smog and cause adverse health effects. Many of the world's megacities, such as Beijing, Cairo, Delhi, Dhaka, Karachi, Mexico City, Shanghai, as well as London and Los Angeles, have considerable air pollution problems. Air pollution has been a serious problem in the world's megacities for decades. Health problems linked to local air pollution, such as lung cancer and chronic respiratory diseases, are a serious problem. These diseases also result in high cost burdens to the world's health systems (WHO, 2000, 2005). About 4 million people die annually prematurely because of outdoor air pollution according to the WHO (2019).

Increased car ownership in emerging economies such as China, India and Mexico is likely to worsen the air pollution problem. Some countries and cities have regulations in place to reduce urban air pollution, for example by (temporarily) closing down polluting industries and introducing a licence plate regulation and restriction system for private vehicles. The city of Beijing is following this approach; however, there are various ways around the system. Innovation in electric vehicles (EV) has partly been fuelled by air pollution concerns. While the use of EVs is increasing, particularly in the EU, China and the US, there is still a lack of a large-scale uptake by users and an absence of abundant charging infrastructure for EVs in many countries. Also, EVs only make sense in terms of mitigating emissions if the electricity they consume comes from renewable or low carbon energy sources. In Sweden and Norway, for example, almost all the electricity comes from low carbon energy sources such as hydropower and nuclear power; therefore, the use of EVs reduces air pollution and reduces stress on limited fossil resources. However, the same cannot be said from China where most of the electricity still comes from coal.

Outdoor air pollution also brings with it the case of transboundary air pollution. *Transboundary air pollution* means that the pollution is caused at one specific geographic location, for example in one city in country A, but due to wind and climatic conditions the pollution is transported to other areas, for example over the border into country B. Transboundary air pollution was a major issue in the 1970s, 1980s and 1990s in relation to acid rain that had its origins in the polluting coal-fired factories of Russia and Eastern Europe, but was swept over to northern and northwestern Europe with the prevailing winds and caused acidification of lakes and water bodies there.

Alternative energy options

The above sections discussed key issues related to energy use and its implications for development as well as the contribution of energy use to climate change, resource depletion and air pollution. The combustion of fossil fuels leads to GHG emissions that cause climate change, the extraction of fossil fuels depletes finite natural resources that have formed over millions of years and the use of fossil fuels contributes to air pollution. So-called low carbon energy technologies, such as renewable energy technology, nuclear energy technology and carbon capture and storage (CCS), are therefore key mechanisms to reducing carbon dioxide and other GHG emissions. Low carbon energy emits less GHGs than conventional fossil fuels such as coal, oil and natural gas. Some low carbon energy technologies, such as large hydropower and nuclear energy, have witnessed a recent revival due to climate change concerns. Nevertheless, this brings with it other adverse effects, such as concerns about health, safety and environmental impacts with regard to nuclear power (Goldemberg & Lucon, 2009).

While about 80% of the world's energy supply comes from fossil fuels, there is an increasing trend towards using alternative non-fossil energy options, such as renewable energy (IEA, 2019). Renewable energy has high growth rates around the world. Due to its rapid implementation time, renewable energy may also avoid carbon lock-in and path dependency. Implementing renewable energy technology today may provide low carbon energy quickly and may avoid lock-in effects, such as dependence on fossil fuel power plants for decades.

Renewable energy comes from renewable natural resources, such as the sunlight, wind, water, tides, geothermal heat and biomass. Unlike fossil fuels and nuclear energy, which are finite and depletable, these energy resources are renewable and non-depletable. Renewable energy has a large global potential. The World Energy Council estimates that the theoretical potential for solar energy is 370 PWh/year, for primary biomass 315 PWh/year, for wind energy 96 PWh/year and for hydropower 41 PWh/year; nevertheless, the technical and economic potential is lower due to variations in land availability and financial competition with fossil fuels (WEC, 2007). About 20% of global electricity consumption came from renewable energy in 2010 and about 25% in 2016, mainly from hydropower, but also from wind, solar and biomass (IEA, 2019). The most widely used and commercialised renewable energy technologies are wind turbines, solar photovoltaic (PV) panels and hydropower technology.

While the environmental benefits of renewable energy are well established, renewable energy also offers an alternative for improving energy access and reducing energy poverty. As we discussed above, the UN's universal modern energy access target by 2030 is expected only to be achievable if a large part of the rural population in developing countries gets access to electricity and clean cooking fuels through renewable energy. This includes options such as mini-grids and off-grid renewable energy, particularly solar and micro-hydro, as well as biogas for cooking (IEA, 2010). This also reduces the reliance on traditional biofuels such as fuelwood, which has adverse health impacts. Nevertheless, there are major barriers, particularly of economic, political and social nature. This book will address these issues throughout the following chapters.

Exercises

1   Go to the statistics website of the International Energy Agency IEA: www.iea. org/statistics. Choose data for the electricity generation of a country of your choice, e.g. China or the UK. This is displayed in GWh (gigawatt hour). Calculate how much the electricity generation is in PWh, TWh, MWh and kWh. See the conversion table at the beginning of the chapter.

2a Go to the IEA Energy Access Database: www.iea.org/energyaccess/database/
    Choose a developing country of your choice, e.g. India or Nigeria. Look at the database to see how many people are still living without access to electricity and without access to clean cooking fuels. How does this relate to the national, urban and rural electrification rates in this country? How does this relate to

number of people relying on traditional biomass? What are the differences and how can this be explained? Take another country as a comparison and explore the same features to understand the dynamics behind these energy poverty issues.

2b   Do some research for your country of choice. Find out how national policies as well as international initiatives such as the Sustainable Development Goal 7 and the Sustainable Energy for All (SE4All) initiative are helping to reduce energy poverty and increase energy access in your chosen country.

3   Go to the NASA website on global climate change: https://climate.nasa.gov/vital-signs/carbon-dioxide/

Look at the most recent data for CO_2 emissions. How does the current level of CO_2 compare to pre-industrial levels and to the 450 ppm target that should not be exceeded to avoid dangerous climate change?

References

Anderson, K. (2009) *Climate Change in a Myopic World*. Tyndall Centre for Climate Change Research. Briefing Note No. 36. Available from: www.tyndall.ac.uk/sites/default/files/bn36.pdf

BBC. (2018) *India Says All Villages Have Electricity*. Available from: www.bbc.com/news/world-asia-india-43946049

Chambers, R. (1995) Poverty and livelihoods: Whose reality counts? *Environment and Urbanization*, 7 (1), 173–204.

Collier, P. (2007) *The Bottom Billion: Why the Poorest Countries are Failing and What Can Be Done About It*. Oxford, Oxford University Press.

Cowen, M. & Shenton, R. (1996) *Doctrines of Development*. London, Routledge.

Cutnell, J.D. & Johnson, K.W. (2012) *Introduction to Physics*. 9th edition. Singapore, John Wiley & Sons.

Goldemberg, J. & Lucon, O. (2009) *Energy, Environment and Development*. 2nd edition. Oxon, Earthscan, Routledge.

Hart, G. (2001) Development critiques in the 1990s: cul de sac and promising paths. *Progress in Human Geography*, 25 (4), 649–658.

Hickey, S. & Mohan, G. (2005) Relocating participation within a radical politics of development. *Development and Change*, 36 (2), 237–262.

Humphrey, J. (2007) Forty years of development research: Transformations and reformations. *IDS Bulletin*, 38, 14–19.

IEA. (2002) *Electricity in India: Providing Power for the Millions*. Paris, International Energy Agency (IEA), OECD/IEA.

IEA. (2010) *World Energy Outlook 2010. Energy Poverty: How to Make Modern Energy Access Universal?* Paris, International Energy Agency (IEA), OECD/IEA. Available from: www.worldenergyoutlook.org/media/weowebsite/2010/weo2010_poverty.pdf

IEA. (2011) *World Energy Outlook 2011. Energy for All. Financing Access for the Poor*. Paris, International Energy Agency (IEA), OECD/IEA. Available from: www.iea.org/media/weowebsite/energydevelopment/weo2011_energy_for_all.pdf

IEA. (2012) *Understanding Energy Challenges in India*. Paris, International Energy Agency (IEA), OECD/IEA. Available from: www.iea.org/publications/freepublications/publication/India_study_FINAL_WEB.pdf

IEA. (2019a) *Statistics*. Paris, International Energy Agency (IEA), OECD/IEA. Available from: www.iea.org/statistics/

IEA (2019b). Energy Access. Paris, International Energy Agency (IEA), OECD/IEA. Available from: www.iea.org/energyaccess/

IPCC. (2007) *Climate Change 2007. Synthesis Report. Contribution of Working Groups I, II and III to the Fourth Assessment Report of the Intergovernmental Panel on Climate Change (IPCC).* [Core Writing Team, Pachauri, R.K and Reisinger, A. (Eds.)]. IPCC, Geneva, Switzerland, 104 pp. Available from: www.ipcc.ch/publications_and_data/publications_ipcc_fourth_assessment_report_synthesis_report.htm

IPCC. (2013) *Climate Change 2013. The Physical Science Basis. Contribution of Working Group I to the Fifth Assessment Report of the Intergovernmental Panel on Climate Change (IPCC).* [Stocker, T.F., D. Qin, G.-K. Plattner, M. Tignor, S.K. Allen, J. Boschung, A. Nauels, Y. Xia, V. Bex & P.M. Midgley (Eds.)]. Cambridge University Press, Cambridge, United Kingdom and New York, NY, USA, 1535 pp. Available from: www.climatechange2013.org/

IPCC. (2014a) *Climate Change 2014. Impacts, Adaptation and Vulnerability. Contribution of Working Group II to the Fifth Assessment Report of the Intergovernmental Panel on Climate Change (IPCC).* [Field, C.B., V.R. Barros, D.J. Dokken, K.J. Mach, M.D. Mastrandrea, T.E. Bilir, M. Chatterjee, K.L. Ebi, Y.O. Estrada, R.C. Genova, B. Girma, E.S. Kissel, A.N. Levy, S. MacCracken, P.R. Mastrandrea, and L.L. White (Eds.)]. Cambridge University Press, Cambridge, United Kingdom and New York, NY, USA. Available from: http://ipcc-wg2.gov/AR5/report/final-drafts/

IPCC. (2014b) *Climate Change 2014. Mitigation of Climate Change. Contribution of Working Group III to the Fifth Assessment Report of the Intergovernmental Panel on Climate Change (IPCC).* [Edenhofer, O., R. Pichs-Madruga, Y. Sokona, E. Farahani, S. Kadner, K. Seyboth, A. Adler, I. Baum, S. Brunner, P. Eickemeier, B. Kriemann, J. Savolainen, S. Schlömer, C. von Stechow, T. Zwickel and J.C. Minx (Eds.)]. Cambridge University Press, Cambridge, United Kingdom and New York, NY, USA. Available from: www.ipcc.ch/report/ar5/wg3/

IPCC. (2018) *Special Report. Global Warming of 1.5°C.* Available from: https://www.ipcc.ch/2018/10/08/summary-for-policymakers-of-ipcc-special-report-on-global-warming-of-1-5c-approved-by-governments/

Jiahua, P., Wuyuan, P., Meng, L., Wu, X., Wan, L., Zerriffi, H., Elias, B. Zhang, C. & Victor, D. (2006) *Rural Electrification in China 1950–2004: Historical Processes and Key Driving Forces.* Program on Energy and Sustainable Development. Stanford University. Working Paper No. 60. Available from: http://iis-db.stanford.edu/pubs/21292/WP_60%2C_Rural_Elec_China.pdf

Jolly, R. (2003) Human development and neo-liberalism: Paradigms compared. In: Kukudar-Parr, S. & Shiva Kumar, A.K. (Eds.) *Readings in Human Development.* New Delhi, Oxford University Press.

Knutti, R., Rogelj, J., Sedláček, J. & Fischer, E.M. (2016) A scientific critique of the two-degree climate change target. *Nature Geoscience,* 9, 13–18.

Leggett, J. (2013) *The Energy of Nations: Risk Blindness and the Road to Renaissance.* Oxon, Earthscan, Routledge.

Mohan, G. & Holland, J. (2001) Human rights and development in Africa: Moral intrusion or empowering opportunity. *Review of African Political Economy,* 88, 177–196.

NASA. (2019) *Climate Change: Carbon Dioxide. Latest measurements: November 2019.* Available from: https://climate.nasa.gov/vital-signs/carbon-dioxide/

OPHI. (2010) *Multidimensional Poverty Index.* Oxford Poverty & Human Development Initiative (OPHI). University of Oxford. Available from: www.ophi.org.uk/wp-content/uploads/OPHI-MPI-Brief.pdf

Practical Action (2010) *The Poor People's Energy Outlook.* Practical Action, Rugby. Available from: http://practicalaction.org/ppeo2010

Pye, S., Watkiss, P., Savage, M. & Blyth, W. (2010) *The Economics of Low Carbon, Climate Resilient Patterns of Growth in Developing Countries: A Review of the Evidence.* Stockholm

Environment Institute (SEI) Report to UK Department for International Development (DFID). Available from: http://sei-international.org/mediamanager/documents/Publications/Climate/economics_low_carbon_growth_report.pdf

Richardson, K., Steffen, W., Schellnhuber, H.J., Alcamo, J., Barker, T., Kammen, D.M., Leemans, R., Liverman, D., Munasinghe, M., Osman-Elasha, B., Stern, N. & Wæver, O. (2009) Synthesis Report. *Climate Change. Global Risks, Challenges and Decisions. 10–12 March 2009, Copenhagen*. University of Copenhagen. Available from: http://climatecongress.ku.dk/pdf/synthesisreport/

Rist, G. (2007) Development as a buzzword. *Development in Practice*, 17 (4–5), 485–491.

Rogelj, J., den Elzen, M., Höhne, N., Fransen, T., Fekete, H., Winkler, H., Schaeffer, R., Sha, F., Riahi, K. & Meinshausen, M. (2016) Paris Agreement climate proposals need a boost to keep warming well below 2 °C. *Nature*, 534, 631–639.

SE4All. (2014) *United Nations Sustainable Energy for All Initiative*. United Nations (UN). Available from: www.se4all.org/

Tans, P. & Keeling, R. (2019) *Recent Monthly Average Mauna Loa CO_2*. NOAA National Oceanic and Atmospheric Administration. Available from: www.esrl.noaa.gov/gmd/ccgg/trends/

World Health Organisation WHO (2019). *Air pollution*. Available from: https://www.who.int/airpollution/en/

UK ERC. (2009) *The Global Oil Depletion Report*. UK Energy Research Centre (ERC). Available from: www.ukerc.ac.uk/support/tiki-index.php?page=Global+Oil+Depletion

UK Government (2000) *Warm Homes and Energy Conservation Act 2000*. UK National Renewable Energy Centre (Narec). Available from: www.legislation.gov.uk/ukpga/2000/31/contents

UNDP. (n.d.) *Human Development Reports*. United Nations Development Programme (UNDP). Available from: http://hdr.undp.org/en

UNDP and WHO. (2009) *The Energy Access Situation in Developing Countries – A Review Focusing on Least Developed Countries and Sub-Saharan Africa*. United Nations Development Programme (UNDP) and the World Health Organization (WHO). Available from: http://content.undp.org/go/cms-service/stream/asset/?asset_id=2205620

Urban, F. (2014) *Low Carbon Transitions for Developing Countries*. Oxon, Earthscan, Routledge.

Urban, F. & Nordensvärd, J. (2013) *Low Carbon Development: Key Issues*. Oxon, Earthscan, Routledge.

Urban, F., Mohan, G. & Zhang, Y. (2011) The understanding and practice of development in China and the European Union. *IDS Working Paper*, 372.

WEC. (2007) *2007 Survey of Energy Resources*. World Energy Council (WEC). Available from: http://minihydro.rse-web.it/Documenti/WEC_2007%20Survey%20of%20Energy%20Resources.pdf

WHO. (2000) *Addressing the Links between Indoor Air Pollution, Household Energy and Human Health*. Geneva, World Health Organization (WHO).

WHO. (2005) *Indoor Air Pollution and Health*. Geneva, World Health Organization (WHO). Factsheet No. 292. Available from: www.who.int/mediacentre/factsheets/fs292/en/

WHO. (2006) *Indoor Air Pollution. Fuel for Life: Household Energy and Health*. Geneva, World Health Organization (WHO). Available from: www.who.int/indoorair/publications/fuelforlife/en/index.html

World Health Organisation (WHO). (2019) *Air pollution*. Available from: https://www.who.int/airpollution/en/

World Bank (2014) *Data*. Washington DC, The World Bank. Available from: http://data.worldbank.org/

World Bank (2019) *Open Data*. Washington DC, The World Bank. Available from: https://data.worldbank.org/

2

ENERGY USE AND ENERGY SYSTEMS IN DIFFERENT COUNTRIES AND CONTEXTS

Energy and energy systems for development: key issues

Key definitions

Energy use refers to the consumption of energy. For example, energy can be used by households or industries, for a wide range of purposes and end-uses.

Energy demand refers to the energy that is demanded or needed by customers. It involves the customer's willingness and ability to pay or access a specific energy service or energy product. For example, demand for electricity is highest during peak times such as in the evenings when lighting is needed. Demand for fuelwood is high in many developing countries, particularly for cooking.

Energy supply is defined as the delivery of a fuel for consumption. To achieve energy supply, several processes can be involved, such as energy extraction, generation, transmission, distribution and storage of fuels. For example, electricity is supplied to grid-connected customers.

The term *energy system* refers to an interrelated network of energy sources, connected by transmission and distribution of that energy to where it is needed. An energy system encompasses the production, generation, transmission and use of energy, including its technology. For example, a grid-based energy system includes the sourcing of energy resources, the power stations or energy technology that generate the electricity, the transmission lines and converter stations that transmit the electricity as well as the end-uses that consume the energy.

Energy carrier refers to a substance or system that contains potential energy that can be released and used as actual energy in the form of mechanical work or heat or to operate chemical and physical processes. Energy carriers include batteries, coal, dammed water, electricity, hydrogen, natural gas, petrol and wood. Energy carriers do not produce energy; however, they 'carry' the energy until it is released.

A *fuel* is a material that stores potential energy that can be released and used as actual energy in the form of heat. Fuels are divided into liquid fuels such as oil or biodiesel, solid fuels such as coal or fuelwood and gaseous fuels such as natural gas or hydrogen. One often refers to fossil fuels, which are fuels that have been formed over millions of years from the organic remains of prehistoric animals and plants. Fossil fuels have a high carbon content and include coal, oil and natural gas. Renewable energy, such as wind, solar and water, are not referred to as fuels. The fuel costs of renewable energy technologies are zero in comparison to the expensive fuel costs of fossil fuels.

An *end-use* refers to the final use of energy or the final activity that is driven by energy. An end-use of electricity is, for example, refrigeration, powering a TV, charging a mobile phone, heating, cooking, boiling water or air conditioning. An end-use of oil-based energy products is, for example, driving a car (petrol/diesel) or flying a plane (kerosene).

Different countries and contexts: impacts on energy use, demand, supply and energy systems

The energy systems, energy use, demand and supply of developed countries are usually characterised by universal modern energy and electricity access, advanced market access, advanced energy infrastructure, advanced energy services and an urban–rural balance in relation to energy access and energy use.

For example, in a Western European country such as the UK, Germany or Sweden, the electrification rate is 100%. Individuals, households and institutions have access to electricity and modern fuels such as natural gas, for a wide range of activities, such as cooking, heating, boiling water, refrigeration, communication, electronic devices, etc. Access to energy markets is guaranteed in most places in the developed world. This means that it is possible to buy equipment, including stoves, radiators, refrigerators, electronics, telecommunication devices, water boilers, kettles, etc., in every city, many towns and certainly online. Similarly, electricity and natural gas is usually delivered right to the doorstep of most households in developed countries. Modern energy infrastructure, such as grid connections or natural gas pipelines, is common. Unless someone lives in a very remote, desolate place in the developed world, they should have access to modern energy equipment, fuels and energy infrastructure on a regular basis. Energy systems and energy use, demand and supply in urban areas are similar to the energy systems and energy use, demand and supply in rural areas in most developed countries. For example, a household in rural Bavaria has similar energy use and enjoys similar energy services based on similar energy systems to a household in Berlin. Similarly, a household in rural Scotland usually has similar energy use, demand and supply and enjoys similar energy services based on similar energy systems (that is usually grid-based electricity from fossil fuels and natural gas for heating/cooking) to a household in London. While energy use per household may differ when economic activities are taken into account, such as when you compare the energy use of a farm with the energy use of a typical non-farm household, the differences between urban and rural areas are rather marginal in most areas of developed countries. The only major difference is for households that are located in remote areas, such as the islands of the Hebrides in Scotland. On small islands such as Eigg and Rum the electricity tends to come from small-scale renewable energy, such as hydropower, wind and solar, rather than from grid-based fossil fuels like in the rest of the UK.

The energy systems, energy use, demand and supply of developing countries are usually characterised by a high use of traditional biomass, low levels of electricity access, limited market access, limited energy infrastructure, limited energy services and an *urban–rural divide* in relation to energy access, energy use, demand and supply.

For example, in many poor countries electrification rates can be very low. Electrification rates in urban areas tend to be higher than in rural areas. Residential electricity use for all of sub-Saharan Africa (excluding South Africa) is roughly equivalent to the electricity use of New York. This means that the 20 million inhabitants of the city use as much electricity as the nearly 800 million inhabitants of sub-Saharan Africa (IEA, 2010). Some Southeast Asian countries, such as Myanmar and Cambodia, also have very low electrification rates. The use of traditional biomass for cooking, lighting, heating and boiling water is therefore high. The IEA estimates that 2.7 billion people worldwide rely on traditional biomass – such as fuelwood, agricultural residues and dung – for basic needs such as cooking and heating (IEA, 2019). In many developing countries there is limited access to energy markets, services and infrastructure. For example, it may not be easy to buy a functioning, modern, electric stove in rural Uganda, or to charge a mobile phone at an electric socket powered by the grid in many parts of rural Cambodia or to find the latest power plant technology in rural Haiti. While there are large differences in the prevalence of energy use, demand, supply and energy systems between rural and urban areas, as well as between more wealthy and poorer groups in these countries, limitations to energy quality and an underperforming power sector affect almost everyone in one way or the other.

Different countries and contexts: economies and their influence on energy use, demand, supply and energy systems

Different types of economies influence the way people live, work, travel, communicate and use energy. The most striking distinctions between the economies of more developed and less developed countries are as follows (Urban et al., 2007; Van Ruijven et al., 2008):

- distinction between formal and informal economies
- distinction between agrarian and industrialised economies
- distinction between more equal and more unequal income distribution

The distinction between *formal* and *informal economies* is important for energy use, demand and supply as the official economic activity (e.g. gross domestic product per capita, GDP/capita) which is often seen as a driving force for energy demand, may not reflect actual economic activity. When economies become formalised by having an increasing income with a declining informal economy, the official economic growth is shown in statistics as artificially high, and *energy intensity* (in gigajoules [GJ] per official dollar) decreases rapidly. This does not reflect the reality, however, as the economic growth is in fact lower; it is only the formalising of economic activities that makes the economic growth appear higher in statistical records (Van Ruijven et al., 2008). It thereby appears that the energy intensity, as well as the *carbon intensity*, declines rapidly, although in reality this is an accounting error rather than an actual improvement.

The distinction between *agrarian* and *industrialised economies* matters as the energy use, demand, supply and the energy systems are very different when a country is predominantly dependent on agriculture to when it predominantly depends on industries. For example, the energy-use patterns and the technologies used for energy generation, transmission and distribution are very different in a country such as Bhutan, which largely relies on subsistence farming, to a country such as Germany, which is heavily industrialised.

The distinction between more equal and more unequal *income distribution* in economies also affects energy use, demand, supply and energy systems. In many OECD countries, the gap between rich and poor is relatively small, although this gap is widening in recent years in some countries. The energy use of a poor European citizen is, of course, different to the energy use of a rich European citizen; nevertheless, the difference will mainly be in the quantities of energy used as the end-uses and the energy sources will be very similar if not the same. In many non-OECD countries, however, there is a large gap between income groups as income is often more unequally distributed. The energy use, demand and supply of a poor Asian or African citizen is therefore likely to be fundamentally different to the energy use, demand and supply of a rich Asian or African citizen. There will be a difference in quantity and quality of energy use and energy systems available. While the wealthier citizen is likely to have access to (more or less reliable) electricity for a range of end-uses such as boiling water, refrigeration, air conditioning, communication and electronics, as well as other energy options such as natural gas, LPG or biogas for cooking, the poorer citizen will likely cook with traditional biofuels such as fuelwood and use fuelwood, candles or kerosene for lighting. Differences will also be apparent in energy use for transport, with richer people being more likely to use petrol, diesel or LPG for private cars, while poorer people are more likely to use LPG for auto-rickshaws and mopeds, at best, and cycle, walk or use animal-driven carriages if they are worse off. The use of public transport may also be more frequent for those lower income groups.

Energy and energy systems in different contexts

Energy and electricity access, markets, services and infrastructure

About 1 billion people worldwide do not have access to electricity and approximately 2.7 billion people rely on traditional biomass – such as fuelwood and dung – for basic needs such as cooking and heating (IEA, 2019). A large part of these people live in South Asia and sub-Saharan Africa and mainly in rural areas (IEA, 2019). Nevertheless, rapid progress has been made in many countries in which electrification rates were very low about one or two decades ago. For example, electrification rates in Afghanistan and Myanmar were around 5% in 2000, whereas today those countries report electrification rates of 98% and 70%, respectively (World Bank, 2018). Good examples for increased access to electricity over several decades are China, Thailand, Malaysia and Vietnam, which have achieved electrification rates of about 100% in recent years, thanks to widespread electrification schemes, large investments and strong government efforts over decades. Despite some very low electrification rates, particularly in sub-Saharan Africa, and about 1 billion people living in darkness, the IEA (2019) suggests that universal access to electricity is possible for everyone, partly due to mini-grid and off-grid renewable energy options in rural areas. This is reflected in the UN Sustainable Energy for All Initiative (SE4All), which aims to provide access to modern energy services to everyone worldwide by 2030 (SE4All, 2014; Urban, 2014).

In the past, it was often assumed that with rising per capita income access to electricity would occur more or less automatically, at an estimated increase of 0.57% per year (IEA, 2002a). Critically assessed, however, it appears to be too simplistic to assume a direct link between income and access to electricity for a number of reasons: Income is not equally distributed, governments can refuse or not have sufficient means to invest in electricity infrastructure (including due to corruption), structural problems may occur as described earlier, technical and geographical limitations exist for grid connections and the rise of violent conflict can lead to a decline in energy infrastructure (as seen in Syria). Further, population growth is also likely to lead to an increase of at least 1 billion inhabitants in developing Asia by 2030 (IEA, 2010). Consequently, the additional 1 billion inhabitants will need to be provided with electricity, which will decrease the annual rise of electrification rates. Stand-alone renewable options or mini-grids are possibilities for electrifying households, but these options are often not included in the national statistics of electrification. Finally, some studies advocate that electrification in developing Asia and Sub-Saharan Africa can only be successful if long-term strategies comprise micro-loans or other forms of financial support for investment in electric equipment (Kanoria, 2006) or if they create opportunities for a higher income generation (Bhattacharyya, 2006), as experience from India has shown. However, the benefits and disadvantages of micro-loans are debated, and some poor people might potentially end up heavily indebted rather than better off due to micro-loans (Urban, 2014). However, what we have seen in recent years is large-scale investment by governments, foreign donors and investors in energy infrastructure, including large mega-dams across developing Asia that have built up energy generation capacity.

Even though electricity consumption is on the rise, the predominant fuels in most parts of the developing world are still traditional biofuels: fuelwood, fuel roots, dung, agricultural waste, crop residues and fuel sticks. Traditional biofuels are primarily used domestically for cooking and heating, particularly in poorer rural areas.

Besides population growth and the unavailability of alternative fuels, changes in the consumption of traditional biofuels are mainly due to three factors:

- changes in income
- the degree of urbanisation
- the degree of industrialisation

The share of traditional biofuels declines in general with rising GDP, although differences exist between countries and regions (Victor & Victor, 2002). When income distribution and geographic distribution are taken into account, the relation between income and traditional biofuel use, particularly fuelwood use, can be different (Victor & Victor, 2002), because traditional biofuels are mainly used by lower-income households and in poorer rural areas (Urban et al., 2007; Urban, 2014). Usually, people in urban areas and in more industrialised economies use lower shares of traditional biofuels.

Many poor countries suffer not only a lack of access to modern energy, but also a lack of access to energy markets, services and infrastructure. This is particularly distinct in rural areas that are often far away from the grid, from cities, trading centres and financial and technological services.

Access to energy markets includes access to energy generation technology (e.g. a fossil fuel power plant, a hydroelectric dam, a solar PV panel, etc.), access to energy transmission and distribution technology (e.g. a grid connection, a mini-grid), access to end-use devices (e.g. an electric cook stove, a biogas stove, an air-conditioner, a TV, a mobile phone) and access to fuels (e.g. electricity, natural gas, coal, oil, LPG, kerosene, charcoal, fuelwood, biogas, etc.). Access to energy markets is often patchy in many areas in poor countries and restricted to basic markets, such as the fuelwood market. In addition, markets generally often tend to be absent or minimal in countries or areas where non-monetary activities and economies prevail, such as subsistence farming, and where energy use and energy systems are restricted to collecting fuelwood and cooking on an open fire (Urban et al., 2007).

Access to energy services includes the delivery of energy in various forms, for example in the form of electricity from the grid or mini-grid, natural gas from a pipeline, biogas from a biogas installation. However, energy services are often limited in poorer countries. In addition, the quality of the energy services tends to be low as blackouts or brownouts (so-called load shedding) are common. This will be discussed in the section 'Examples of threats to energy security'.

Access to energy infrastructure includes access to power plants, grid lines or pipelines and infrastructure at the interface of energy supply and energy use such as electric sockets, petrol stations or charging stations for electric vehicles. Again, there is a fundamental lack of modern energy infrastructure and energy choices available in poor countries and poor areas.

Energy quality and the performance of the power sector

Energy quality is as important as energy quantity. One of the key issues related to energy quality is energy security. *Energy security* is defined as the availability of energy supply at adequate prices, in adequate quantities and at adequate times to such an extent that the social and economic development of a country can be ensured. Energy security refers both to technical security of supply and quality of supply (Kowalski & Vilogorac, 2008).

Many poorer countries have time limitations for power availability, with electric power only being available for several hours a day (e.g. from 6–9 am and again from 6–10 pm). Private firms, private organisations and richer households may use a private generator as an alternative to electricity from the low-quality grid. Many poorer countries also have inadequate prices; for example, Cambodia has average electricity prices that are similar to the electricity prices in the European Union, despite people living on a very low income. In addition, many poorer countries suffer from an inadequate supply of electricity and energy in general, with a power under-capacity, as not enough electric generating power is available or not enough fuels (e.g. petrol, diesel) are being imported and available for use.

Examples of threats to energy security

Some countries are almost entirely dependent on energy imports. Many less developed, resource-poor countries are deprived of both infrastructure and investment, which has negative implications for the power sector. Countries that are poor in fossil fuel resources often depend on foreign energy imports that make them vulnerable to shocks and trade disputes. Some countries that are fossil fuel resource poor, but have significant water resources, may choose to develop their hydropower potential by building dams in an effort to improve their energy security (provided they find investors to invest in and develop the dams). This is a pathway that has been chosen by several countries in the Greater Mekong sub-region, such as Cambodia, Laos and Myanmar. Other renewable energy options, such as solar, wind and modern biomass, may also help to overcome bottlenecks in energy security. Whatever strategy a country chooses to take, it is *important to diversify the energy mix* by using several energy sources for guaranteeing energy security.

Many developing countries suffer from a *poor performance of the power sector* for various supply-side, demand-side and economic reasons. We discussed the demand-side problems in relation to a lack of access to electricity and the predominant use of fuelwood. The supply-side problems are elaborated below.

From a supply-side perspective, the power system configurations are often sub-optimal in poorer countries: Energy systems are often not meeting the demand, even though a substantial excess capacity may exist (Schramm, 1990). The reasons for the excess capacity are faulty planning and poorly performed operational and maintenance tasks leading to frequent plant breakdown, outages and voltage fluctuations. This results in an unreliable service and causes economic losses (Schramm, 1990; Urban, 2014).

For modern economies to function, it is crucial to meet the electricity demand. Supply shortages are unfortunately characteristic for many developing countries and regions. They can be caused by:

a   a poor performance of the power sector such as poor conditions of generation and distribution equipment, inadequate operational and maintenance performance and a high level of technical and nontechnical losses

b   a rapidly growing demand for electricity;

c   a low number of power plants;

d   technical constraints;

e   organisational and institutional problems;

f   underfinanced power companies;

g   restriction on capital available for investments;

h   a dependence on import of plants and equipment for power supply; and

i   too low consumer prices.

See Van der Werff and Benders (1987), Schramm (1990), IEA (2002b), Urban et al. (2007) and Urban (2014).

A reliable electricity supply can in some situations also be threatened by high electric power transmission and distribution losses, such as in Haiti where they accounted for 55% of the electricity supply in 2011 and the Democratic Republic Congo (DRC) where they accounted for 46% in 2011 (World Bank, 2019). Iraq and Nepal also have high electric power transmission and distribution losses of about 35%, with the Dominican Republic, Cambodia, Namibia and Yemen also having losses of about 30% in 2011. Some Eastern European countries, such as Albania, Montenegro, Serbia and recently Lithuania have high transmission and distribution losses of around 20% in 2011, which is similar to the situation in India (World Bank, 2019). This compares to a rather low Western European transmission and distribution loss rate of about 5% (Urban et al., 2007; World Bank, 2019; Urban, 2014).

From an economic perspective, poor sector financing is common with tariffs below long-term marginal costs of production or even below average operating costs and a poor revenue collection performance by the utilities where a large share of the non-paid bills will never be collected (Van der Werff & Benders, 1987; Schramm, 1990; IEA, 2002b). This is mainly because governmental departments and government-owned companies are under government protection and therefore frequently do not pay their electricity bills, or because customers are simply too poor to pay their bills (Schramm, 1990; IEA, 2002b). Another reason for the financial deficiencies is the common phenomena of electricity theft. India's power sector, for instance, is facing a serious risk of bankruptcy, mainly because of unpaid bills and high transmission and distribution losses. This problem is to some extent related to electricity theft, although the exact amount of electricity stolen is difficult to quantify (IEA, 2002b; Urban et al., 2007; Urban, 2014).

The energy trilemma

The World Energy Council (WEC) designed a tool to rank countries according to their ability to provide adequate energy security, energy equity (accessibility and affordability) and environmental sustainability in their energy provision. This tool assesses the current policies in place, as well as how the trade-offs between these three indicators are being handled. In 2018, Denmark, Switzerland and Sweden ranked at the top three of the Energy Trilemma list (WEC, 2019). At the very bottom of the list were primarily countries in sub-Saharan Africa, such as Niger, Chad, Benin, Democratic Republic of Congo and Tanzania (WEC, 2019). You can see the ranking at https://trilemma.worldenergy.org/.

Different contexts

This section discusses differences of energy use, demand, supply and energy systems between and within different countries and regions. There are considerable differences and variations between different types of developing countries and different groups in society. The most notable differences are between urban and rural areas, the so-called **urban–rural divide**, **different income groups**, different world regions such as **Africa, Asia and Latin America** as well as *emerging economies* such as China and India.

Electricity use is especially high in North America, Europe, Japan, Australia, as well as in and other wealthy countries (e.g., wealthy Arab countries). Electricity use is also high in urban areas in China (particularly the coastal areas in China's East where large, relatively well-off urban areas are situated), urban areas in India and urban areas in other middle-income countries. Large parts of sub-Saharan Africa, parts of Latin America and parts of developing Asia have far lower electricity use.

As mentioned in Chapter 1, there are two so-called *hotspots of energy poverty*: one in sub-Saharan Africa and one in developing Asia. India currently has an electrification rate of about 93% after large-scale government efforts; nevertheless, many rural dwellers still have no access to electricity (IEA, 2019).

Another emerging economy, China, achieved an electrification rate of almost 100% in recent years due to a decade-long history of providing rural electricity access through small-scale hydropower and recent large-scale government-funded electrification programmes. The large majority of the rural population has access to electricity in China and virtually 100% of the urban population (World Bank, 2019).

Many other emerging economies are generally doing well in terms of electricity access and reducing the use of traditional biomass for cooking, particularly in Latin America. Argentina, Brazil, Mexico and Venezuela have all electrification rates between 95% and almost 100% on average (IEA, 2019). Thailand, Malaysia and Indonesia are also doing well with an electrification rate of 100%, 100% and about 98% (respectively), whereas Pakistan's electrification rate is 71% (lower than

India on average). South Africa has also an average electrification rate that is slightly lower, namely around 85%. These average figures, however, conceal the differences between rural and urban areas (IEA, 2019). Despite the increasing electricity access in some rural areas, there are still millions of rural people that do not have access to electricity and cook using traditional biofuels. This shows the problem of the rural-urban divide concerning modern energy access.

Another issue is the divide in energy use, demand, supply and energy systems in different income groups. The wealthy in many developing countries consume energy in similar patterns as the well-off in developed countries. They consume high quantities of efficient, high-quality energy (e.g. electricity for lighting, petrol or diesel for transport, natural gas or LPG for cooking) and have similar end-uses (e.g. air conditioning, TV, mobile phones, heating, cooking, driving, etc.) to their counterparts in richer countries. With rising income, the level of disposable income rises and hence energy use usually rises with it. The poor in many developing countries use far less energy. They often use low-quality, inefficient energy (such as fuelwood and dung) and have limited end-uses (often mainly cooking and water heating). Averaged data on (national) energy demand therefore tends to hide extreme wealth and extreme poverty, rather suggesting averaged energy demand.

Energy-use patterns become more complex, however, once the issue of the **urban poor** and the **rural rich** is investigated. The urban poor often have better access to more energy-efficient and high-quality energy than the rural rich. The problem of the urban poor is that they cannot afford to pay for these high-quality fuels. The problem of the rural rich is that energy-efficient, high-quality fuels are often difficult to access and may involve considerable time and transaction costs. A study from the rural rich in Pakistan argues that the rural rich rather prefer to cook with fuelwood (and thereby be considered to live in energy poverty), which is easily accessible, than to cook with LPG, which is not locally available and is difficult to access in rural areas (Mirza & Kemp, 2011).

Economies and their influence on energy use, demand, supply and energy systems

The formal versus informal economy

Informal economies (also called shadow economies) are characterised by unofficial, often unrecorded transactions that are not recorded in official economic descriptions (Van Ruijven et al., 2008). There is no internationally agreed definition, but usually the definition includes unofficial monetary and non-monetary transactions (e.g. trade or exchange of goods) and may also include illegal activities and tax evasion (e.g. black labour).

Informal economies tend to be driven by tax burdens and social security contributions, overregulation of labour markets and high costs of official labour markets (Schneider & Enste, 2000; Van Ruijven et al., 2008).

In contrary, *formal economies* are characterised by official monetary transactions that are recorded in official economic descriptions such as GDP or GNI and that are taxed.

The informal economies exist everywhere (e.g. the cleaner in the USA, the small-holder farmer in Bolivia or the street stall owner in Bangkok who do not pay taxes or receive social security contributions); nevertheless, the prevalence of the informal economy is much higher in developing countries. Schneider (2005) found that on average the informal economy was about 40% of the official GDP of developing countries in 1999/2000, whereas in OECD countries it was only about 17% on average (Van Ruijven et al., 2008).

It is difficult to measure information on informal economies, as often it has to be derived from indirect indicators. Direct methods to estimate the informal economy include surveys and samples, yet they might not be representative for the entire economy. Another method is to analyse the differences between various statistics, such as between national expenditures and national income data. More advanced approaches include a complex set of indicators that aim to analyse real economic activity, including the amount of transactions in an economy and the physical resource inputs in an economy (e.g. electricity) (Van Ruijven et al., 2008).

As briefly discussed above, the difference between the formal and the informal economy matters for energy use. While formal economies report the large majority of economic activities and thereby have an accurate depiction of the relationship between their GDP and energy use, informal economies tend to have a distorted relationship between their GDP and their energy use, meaning that the energy intensity and the carbon intensity are lower in statistical records than they are in reality.

The size of the informal economy usually declines with increasing income. When informal activities are formalised, the official economic growth appears artificially high and energy intensity, measured as energy use per economic unit, decreases rapidly. This means that energy efficiency figures may be overestimated (Van Ruijven et al., 2008).

This is reflected in Figure 2.1, which shows the rapidly declining energy intensity of the Chinese economy. This may in fact not only be due to rapid improvements in energy efficiency and decarbonisation strategies, but due to a rapid formalisation and monetarisation of the economy. This means that the actual economic activity may be higher; hence, the relationship between energy demand and economic activity may be underestimated.

Equal versus unequal income distribution

Global wealth is unequally distributed. It is being reported that the richest 10% of the world have as much income as the bottom 90%. Global assets are even more unequally distributed with the richest 1% owning 40% of global assets and predominantly being of US, European or Japanese origin (BBC, 2014).

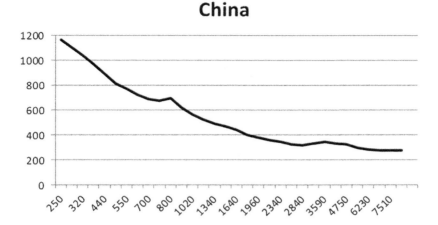

China

FIGURE 2.1   Rapidly declining energy intensity of the Chinese economy: energy intensity versus GDP/capita for China from 1980 to 2011

Source: Data from World Bank.

Note: The x-axis shows GDP/capita (PPP) in 2013 US$, the y-axis shows energy intensity in kilograms of oil equivalent per US$1000 GDP (constant 2005 PPP).

On the other hand, national wealth is also unequally distributed in many countries. The economies of developed and developing countries further differ from each other by differences in *income distribution*. The income distribution describes the range of incomes over the entire population of a country.

While income distribution tends to be more equal in developed countries, income distribution tends to be less equal in developing countries. Particularly the Nordic countries such as Norway, Sweden, Finland, Denmark and Iceland have rather equal income distributions. Of course, there are differences in income and there is a divide between the poor and the rich, as in every country. Nevertheless, the divide between the poor and the rich is smaller in developed countries. This is partly due to social security provisions that help those in need, such as unemployment benefits, maternity and paternity pay, pensions and healthcare.

A recent trend in OECD countries, however, has been an increasing gap between the rich and the poor and thereby a trend towards more unequal income distribution in developed countries. In recent years the global economic crisis has put pressure on economies, income and labour markets all over the world. This has led to the increased accumulation of wealth by the wealthy, a thinning out of the middle class and a larger group of poorer people that tend to be even more disadvantaged than a few years ago.

Developing countries, particularly middle-income countries such as India and/or resource-rich countries such as Angola and Venezuela, tend to have more unequal income distribution, dividing societies between rich elites and poor masses

(Van Ruijven et al., 2008). The economist Simon Kuznets developed the concept of the so-called *Kuznets curve* that indicates that the development of an economy follows an inverted U-shaped development where income inequality first increases with increasing GDP levels and after a threshold the income inequality declines with further increasing GDP levels (Kuznets, 1955; Goldemberg & Lucon, 2009). The Kuznets curve is highly debated in academic literature, with scholars affirming, rejecting and fiercely discussing the concept and the phenomenon of increasing and decreasing income distribution with the development of a national economy (see e.g. Saith, 1983; Glomm, 1997; Deininger & Squire, 1998; Fields, 2001).

Recent criticism of the Kuznets curve is partly based on the rapid economic development of East Asian countries, such as Japan and the so-called *Asian tiger economies* of Hong Kong, Singapore, South Korea and Taiwan between 1965 and 1990. With still relatively low income levels, inequality decreased rapidly. This was contrary to Kuznets' theory that suggested that rapid economic growth would lead to an increase in inequality in the beginning. The economist Joseph Stiglitz argues that the East Asian economic and social development defies the concept of the Kuznets curve due to the following reasons: These states rapidly reinvested the income in long-term policies and programmes that reduced income inequality such as universal education, enacted industrial policies that increased wages and limited the prices of commodities and land reform that increased rural productivity and income. As economic growth occurred, inequality was reduced due to the governments' long-term interventions and reinvestments (Stiglitz, 1996). Despite this critique of the Kuznets curve, it has been observed that non-OECD countries have a higher variation in income distribution (and have therefore more unequal income distributions) compared to OECD countries (see e.g. Van Ruijven et al., 2008).

For measuring inequality the so-called *Gini coefficient* is used. 'The Gini coefficient measures how equal or unequal societies are in terms of their income distribution. The Gini coefficient is given as a percentage and has values between zero (perfect equality) and 100 (perfect inequality)' (Van Ruijven et al., 2008: 2818).

We will come back to the Kuznets curve in Chapter 4, when we discuss the concept of the *environmental Kuznets curve*.

The income distribution of a country influences energy use, demand supply and energy systems

In an equal society where most people have equal incomes and equal lifestyles, energy use, demand and supply tend to be similar in terms of the quantities, quality and types of energy used as well as the end-uses. Similarly, energy systems are rather similar as one predominant system prevails, for example the existence of a stable, reliable central grid with electricity being generated from modern fossil fuel plants.

In an unequal society where large differences exist between the poor and the rich in terms of their incomes and lifestyles, their energy use, demand and supply will also differ significantly. They will differ in terms of the quantities of energy used (e.g. a wealthy person will use a lot of energy for air conditioning and lighting, whereas a poor person will not have any air conditioning and electric lighting at all). There are also differences in the quality of energy (e.g. wealthy households may own private generators that cover up power outages, whereas poor households may be constantly affected by power outages or may not even be connected to the grid), the types of energy used (e.g. fuelwood for cooking versus natural gas or electricity for cooking) and differences in end-uses (e.g. air conditioning, TVs, stereos, video games, mobile phones and iPads versus a fan or radio). Similarly, energy systems will be rather different as the wealthy may depend on grid-based or generator-based energy systems whereas the poor may depend on fuelwood and open fires as an 'energy system'. In developing countries, there are several dominant energy systems, and members of the middle class often use several fuels for different end-uses, such as electricity for lighting and fuelwood for cooking.

Energy demand is usually modelled as a function of average GDP per capita. Changes in income distribution are often not accounted for. Research suggests that income distribution can be an important issue for determining energy demand patterns and fuel choices in the electricity and transport sectors (Andrich et al., 2013).

The energy demand and energy use for low- and high-income groups can therefore be very different from the average energy demand and energy use that is usually calculated at the national level (Van Ruijven et al., 2008). For example, for high-income groups the transport energy demand may be underestimated, whereas for low-income groups the use of traditional biofuels may be underestimated.

Agrarian versus industrialised economies

There are large differences between the structure of economies between developed and developing countries. In high-income countries, agriculture accounted for only about 1% of the GDP in 2010, on average, while the large majority of the economy was based on services (74%) and industry (25%). In low- and middle-income countries, agriculture accounted, on average, for 10% of GDP, industry for about 35% and services for about 55% in 2010. In low-income countries alone, agriculture accounted for almost 30% of GDP, industry for approximately 20% and services for almost 50% in 2010 (World Bank, 2014).

For a predominantly rural agrarian economy in non-OECD countries that depend largely on agriculture, including subsistence farming and small-holder farming, energy use may be rather low. Energy will be primarily needed for farming activities, such as electricity for irrigation and diesel for powering agricultural machines such as tractors. Nevertheless, in many poor countries agriculture is still hardly mechanised and many agricultural tasks are still performed by human and animal power, meaning that energy use is even lower. Countries that have large-scale intensive agricultural systems such as Spain, the Netherlands and the USA consume much more energy. Nevertheless, even these countries have only a small share of GDP value added that comes from agriculture (2.7% for Spain, 2.0% for the Netherlands in 2010 and 1.2% for the USA) (World Bank, 2014).

For industrialised economies, energy use, demand and supply are different and considerably higher. Industries require energy-intensive inputs, large amounts of fuels and energy resources as well as a more advanced energy supply system. One can differentiate between heavy and light industry, with heavy industry requiring more energy and usually having higher greenhouse gas emissions.

Economies that are predominantly service based have lower energy requirements than predominantly industry-based economies; nevertheless, this depends on the size and scale of the service economy. Service economies that are based on small-sized offices and energy-extensive services have much lower energy use and energy demand than service economies that are based on large-scale service buildings and energy-intensive services. For example, a pedicab business or a street stall in Asia has hardly the same energy requirements as a taxi business or a large luxury restaurant in Europe.

Service-based and industry-based economies also tend to have a higher degree of urbanisation. The energy use of the urban population is usually (much) higher than the energy use of the rural population in non-OECD countries. Urban areas have larger middle classes, more energy-intensive lifestyles and greater consumer patterns.

During the process of economic development, it has been observed that *economic restructuring* occurs. This involves the shifting of value added and employment from one sector to another sector. Typically, less developed economies are based on a large share of the population employed in the agricultural sector and a large share of the GDP coming from agriculture. This typically changes with increasing industrialisation as economies become more developed. In a later stage, the employment and share of GDP from the service sector increases. This has been observed in today's industrialised countries during the Industrial Revolution and more recently also in East Asia, particularly in the Asian tiger economies of Hong Kong, Singapore, South Korea and Taiwan between 1965 and 1990. Today China and India show similar levels of development. Nevertheless, economic restructuring can be questioned as it is a highly stylised concept and there may be limitations to it in reality (Van Ruijven et al., 2008; Urban, 2010). Figure 2.2 shows the concept of economic restructuring. The graphs in Figures 2.3 and 2.4 show how economic restructuring took place in China and India during recent economic development.

The historically observed decline of the industrialised sector in OECD countries can partly be explained by the outsourcing and offshoring of industry to low- and middle-income countries. This is one way that energy use and CO_2 emissions could be decoupled from economic growth. The issue of decoupling, however, is fiercely debated. *Decoupling* of energy use and emissions from economic growth ultimately leads to a *dematerialisation* of the economy which leads to sectoral changes from energy- and material-intensive sectors, such as industry, to energy- and material-extensive sectors, such as services. The concept of dematerialisation implies that the amount of materials used – and associated energy and emissions – declines with increasing national income. This is partly due to technological improvements, policies and sectoral changes driven, for example, by outsourcing and offshoring. While

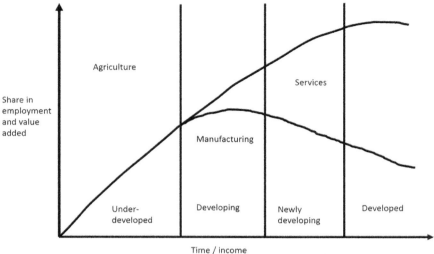

FIGURE 2.2   Economic restructuring from agriculture to industry (manufacturing) to services

Source: Adapted from Van Ruijven et al. (2008: 2812)

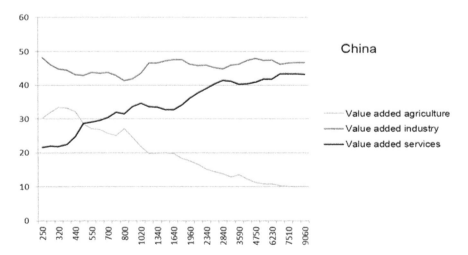

FIGURE 2.3   Value added versus GDP/capita for China from 1980 to 2011

Source: Data from the World Bank.

Note: The *x*-axis shows GDP/capita (PPP) in 2013 US$, the *y*-axis shows value added in %.

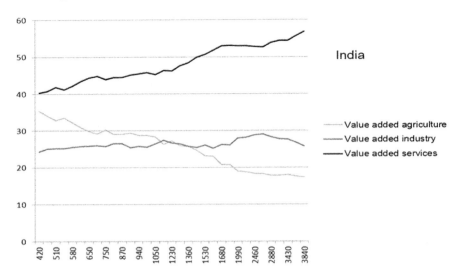

FIGURE 2.4 Value added versus GDP/capita for India from 1980 to 2011

Source: Data from the World Bank.

Note: The *x*-axis shows GDP/capita (PPP) in 2013 US$, the *y*-axis shows value added in %.

OECD countries, particularly in Europe and North America, have outsourced and offshored their energy use and emissions to China, Cambodia, Bangladesh and other low- and middle-income countries, at a global level the concept of dematerialisation remains tricky. This is a race to the bottom, as such offshoring can obviously not be reproduced by countries that currently have the lowest income levels (Van Ruijven et al., 2008; Urban, 2010).

Exercises

1 Choose two countries of your choice: one high-income country and one low/middle-income country (e.g. Germany and Ghana). Go to the World Bank database to find some data to support your research: https://data.worldbank.org/

   1a Discuss the following three issues for these two countries and explain how and why these issues differ between the high-income and the low/middle-income country:

   • distinction between agrarian and industrialised economies (e.g. using value-added data)
   • distinction between formal and informal economies (e.g. using formal/informal employment data)
   • distinction between more equal and more unequal income distribution (e.g. using Gini coefficient data)

1b What does this mean for the energy sector of your two chosen countries? How do these issues impact energy supply and demand, as well as the reporting of those two?

2 Choose a country of your choice and do some research to better understand their energy security situation. You may wish to use some data from the World Energy Council's Energy Trilemma ranking: https://trilemma.worldenergy. org/ as well as information from the IEA and the World Bank.

How can the country of your choice provide energy supply at adequate prices, in adequate quantities and at adequate times to such an extent that the social and economic development of the country can be ensured? How dependent is the country on imports of energy and how independent is it from external actors in terms of using domestic energy resources?

References

Andrich, M.A., Imberger, J. & Oxburgh, E.R. (2013) Inequality as an obstacle to sustainable electricity and transport use. *Energy for Sustainable Development*, 17 (4), 315–325.

BBC. (2014) *Key Facts: The Global Economy Rich and Poor*. BBC News. Available from: http://news.bbc.co.uk/1/shared/spl/hi/guides/457000/457022/html/nn5page1.stm

Bhattacharyya, S.C. (2006) Renewable energy and the poor: Niche or nexus? *Energy Policy*, 34 (6), 659–663.

Deininger, K. & Squire, L. (1998) New ways of looking at old issues: Inequality and growth. *Journal of Development Economics*, 57 (2), 259–287.

Fields, G. (2001) *Distribution and Development, A New Look at the Developing World*. New York, Russell Sage Foundation, and Cambridge, MA, and London, The MIT Press.

Glomm, G. (1997) Whatever happened to the Kuznets curve? Is it really upside down? *Journal of Income Distribution*, 7 (1), 63–87.

IEA. (2002a) *World Energy Outlook 2002. Energy and Poverty*. Paris, International Energy Agency (IEA), OECD/IEA.

IEA. (2002b) *Electricity in India: Providing Power for the Millions*. Paris, International Energy Agency (IEA), OECD/IEA.

IEA. (2010) *World Energy Outlook 2010. Energy Poverty: How to Make Modern Energy Access Universal?* Paris, International Energy Agency (IEA), OECD/IEA. Available from: www. worldenergyoutlook.org/media/weowebsite/2010/weo2010_poverty.pdf

IEA (2019) *Energy Access*. Paris, International Energy Agency (IEA), OECD/IEA. Available from: www.iea.org/energyaccess

Kanoria, H. (2006) Financing the end users of solar energy systems. In: Sastry, E.V.R. & Reddy, D.N. (Eds.) *International Congress on Renewable Energy ICORE 2006 Proceedings*. Hyderabad, Allied Publishers Pvt. Ltd. pp. 186–190.

Kowalski, G. & Vilogorac, S. (2008) *Energy Security Risks and Risk Mitigation: An Overview*. UNECE Annual Report 2008. pp. 77–83. Available from: www.preventionweb.net/files/8066_Pagesfromannualreport2008.pdf

Kuznets, S. (1955) Economic growth and income inequality. *The American Economic Review*, 45 (1), 1–28.

Mirza, B. & Kemp, R. (2011) Why the rich remain energy poor. Consilience. *The Journal of Sustainable Development*, 6 (1), 133–155. Available from: www.consiliencejournal.org/index.php/consilience/article/viewFile/180/69

Saith, A. (1983) Development and distribution: A critique of the cross-country U-hypothesis. *Journal of Development Economics*, 13 (3), 367–382.

Schneider, F. & Enste, D.H. (2000) Shadow economies: Size, causes and consequences. *Journal of Economic Literature*, 38, 77–114.

Schneider, F. (2005) Shadow economies around the world: What do we really know? *European Journal of Political Economy*, 21, 598–642.

Schramm, G. (1990) Electric power in developing countries: Status, problems, prospects. *Annual Review of Energy*, 15, 307–333.

SE4All. (2014) *United Nations Sustainable Energy for All Initiative*. United Nations (UN). Available from: www.se4all.org/

Stiglitz, J.E. (1996) Some lessons from the East Asian miracle. *World Bank Research Observer*, 11 (2), 151–177.

Urban, F. (2010) Pro-poor low carbon development and the role of growth. *International Journal of Green Economics*, 4 (1), 82–93.

Urban, F. (2014) *Low Carbon Transitions for Developing Countries*. Oxon, Earthscan, Routledge.

Urban, F., Benders, R.M.J. & Moll, H.C. (2007) Modelling energy systems for developing countries. *Energy Policy*, 35 (6), 3473–3482.

Van der Werff, R. & Benders, R.M.J. (1987) *Power Planning in India. Adjustment of a Simulation Model to the Electricity Supply System in the State Maharashtra*. IVEM, University of Groningen, Groningen. Work Report No. 6.

Van Ruijven, B., Urban, F., Benders, R.M.J., Moll, H.C., Van der Sluijs, J., De Vries, B. & Van Vuuren, D.P. (2008) Modeling energy and development: An evaluation of models and concepts. *World Development*, 36 (12), 2801–2821.

Victor, N.M. & Victor, D.G. (2002) *Macro Patterns in the Use of Traditional Biomass Fuels*. Program on Energy and Sustainable Development. Stanford University, Stanford.

World Bank (2018) *Data*. Washington DC, The World Bank. Available from: http://data.worldbank.org/

World Bank (2019). *Open data*. Available from: https://data.worldbank.org/

World Energy Council (WEC) (2019) *The Energy Trilemma*. Available from: https://trilemma.worldenergy.org/

3

ENERGY TRANSITIONS

From traditional biomass to fossil fuels to low carbon energy

The concept of energy transitions

Transition is here defined as a system change, rather than a neatly organised or orchestrated change.

Energy transitions are shifts from a country's economic activities based on one energy source to an economy based (partially) on another energy source. Fouquet and Pearson (2012) find that energy transitions are complex and rare events that have often taken place over very long periods, such as decades or centuries.

Several energy transitions have occurred in history:

- the energy transition from human power and animal power to traditional biomass (such as fuelwood, crop residues and dung)
- the energy transition from traditional biomass to coal (ca 1860)
- the energy transition from coal to oil (ca 1880)
- the energy transition from oil to natural gas (ca 1900)
- the energy transition from natural gas to electricity and heat (ca 1900–10)
- the large-scale commercial introduction of nuclear (ca 1965)
- the large-scale commercial introduction of renewable energy and large hydropower (ca 1995)

(Bashmakov, 2007)

Nevertheless, these energy transitions are incomplete, with several energy carriers being used at the same time, such as natural gas for cooking and electricity for lighting. These observations are mainly based on developments in developed countries. In many developing countries, energy transitions are more dependent on income class than on discrete and time-wise defined transitions (see also Sovacool, 2011). In developed countries these transitions occurred at different paces too; for example, the energy transitions from traditional biomass to coal to oil were much faster in the Unites States than in the United Kingdom. This was due to several factors such as resource availability, government policy, industrial and household energy demands (Fouquet & Pearson, 2012), as well as energy prices (Fouquet, 2016).

Energy transitions are characterised by *changing patterns of energy use* (e.g. from solid to liquid to electricity), *changing energy quantities* (from scarcity to abundance or the other way around) and *changing energy qualities* (e.g. from fuelwood to electricity) (Bashmakov, 2007).

Bashmakov's *three laws of energy transitions* suggest the following:

a   Energy transitions are often driven by *changing energy costs in relation to income* (the predominant energy form becomes too expensive).
b   Energy transitions are often driven by *improving energy quality* (e.g. higher energy efficiency like electricity in comparison to fuelwood).
c   Energy transitions are often driven by *growing energy productivity* (e.g. more industrial output can be obtained).

Figure 3.1 shows the historically observed energy transitions between 1980 and 2010.

Grubler (2012) analysed the period of time it took for historical energy transitions. He concludes that most of the transitions took place within a time frame

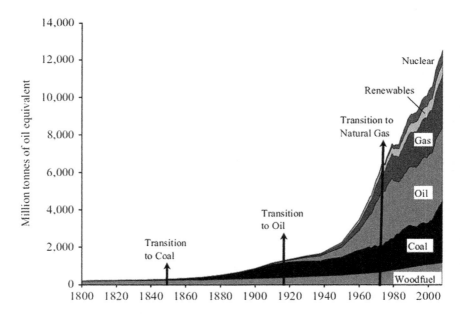

FIGURE 3.1   Global energy consumption and energy transitions between 1800 and 2010

Source: Fouquet and Pearson (2012: 2)

of about 80–130 years for fossil fuel carriers like coal, oil and natural gas, whereas traditional biomass has been used for more than 2000 years. It needs to be mentioned that none of these energy transitions has been complete as all of these energy carriers are still being used worldwide. Yet, the observed trend is that as economies become more developed and households have higher incomes, energy transitions tend to happen from traditional biomass to fossil fuels. In the long run, low carbon energy transitions may be possible.

Low carbon energy transitions

In the light of climate change, recent research has been conducted to assess how low carbon energy transitions can take place. *Low carbon energy transitions* can be defined as shifts from a country's economic activities based on fossil fuels to an economy based (partially) on renewable and low carbon energy. This means that substitutions take place from fossil-fuel-based technologies to low carbon technologies. Such transitions can take place in every sector of a country's or a region's economy. Low carbon energy transitions are likely to open up new opportunities for developing countries to reduce poverty, achieve higher development and higher living standards such as by providing modern energy access through renewable energy, while at the same time mitigating climate change and safeguarding fossil energy resources (Urban, 2014).

Low carbon energy transitions can be achieved by (a) *reducing energy use*, (b) *introducing renewable and low carbon energy*, (c) increasing *energy efficiency* and using fossil fuels efficiently, as Lysen (1996) described in the concept of the *Trias Energetica*. Nevertheless, this view is rather *technocentric* and neglects the complex interactions between policy and technology as well as *socioeconomic factors* that influence energy transitions.

Observations from history have shown that most energy transitions were rather slow, often lasting more than a century (Solomon & Krishna, 2011; Pearson & Foxon, 2012). Faced by a changing climate and fossil fuel depletion, many scholars suggest that a low carbon energy transition needs to be more rapid. A few low carbon transitions were successful, such as Brazil's transition from an oil-based transport system to a sugarcane-based bioethanol transport system and France's transition from predominantly oil-fired electric power to nuclear power (Solomon & Krishna, 2011). However, other attempts have failed, and it is not entirely clear how a low carbon energy transition can be spurred. Grubler (2012) and Fouquet and Pearson (2012) note that low carbon innovation, policies and financing need to be more persistent, continuous and balanced on a national and global level. They suggest that current policy frameworks are insufficient and need to be changed for enabling a low carbon energy transition.

Nevertheless, achieving a low carbon energy transition will not only require new technology, adequate policies and financing, but a systems transition that requires a fundamental change in practices, behaviours and politics. For example, shifting from oil-based cars to electric cars will not only require a change in car innovation and technology, but it will require a completely different infrastructure with electric charging stations and a radical change in consumer behaviour and individual and mass preferences. There is a wide range of literature on systems transitions and so-called *sociotechnical transitions* for low carbon innovation. Sociotechnical transformations require a change to social practices, norms, infrastructures, techno-scientific knowledge, networks and symbolic meanings (e.g. Geels, 2002, 2005).

Urban and Nordensvard (2018) analyse low carbon energy transitions in the Nordic countries between 1960s and 2018. They find that low carbon energy transitions have occurred over the last decades in Denmark, Iceland, Finland, Norway and Sweden. For example, Denmark shifted its reliance from fossil fuels to wind energy; Iceland generates virtually all its electricity and heat from geothermal energy and hydropower; Norway's electricity comes almost completely from hydropower and its heat comes to a large extent from biofuels; Sweden's electricity mix is almost completely carbon-free due to its reliance on hydropower, nuclear, wind and biofuels, while its heat is mainly generated from biofuels; and Finland has also introduced a larger share of hydropower, nuclear power and biofuels in its electricity mix (IEA, 2018). Urban and Nordensvard (2018) found that the Nordic countries are therefore evidence that low carbon energy transitions are possible, while at the same time increasing and maintaining high levels of national prosperity and individual well-being. Unlike many other countries in the world, there is evidence that between 1960 and 2018, total CO_2 emissions have declined in Denmark, Iceland and Sweden, while they declined in Finland between 1990 and 2018. This is mainly due to the wide-ranging energy transitions implemented in these countries. Per capita CO_2 emissions have also declined between 1960 and 2018 in Denmark, Finland, Iceland and Sweden, while GDP and GDP PPP rose almost exponentially for these countries during the same time (World Bank, 2018). For example, for Sweden GDP rose by about 35 times between 1960 and 2018, yet emission levels are currently about 15% below 1960 levels. This was not always the case. Indeed in the late 1960s and 1970s, CO_2 emissions were extremely high (almost a doubling of emissions happened between 1960 and 1970), yet the oil crisis in the 1970s let to a large-scale restructuring of the energy mix and a breaking free from oil and other fossil fuel dependency. Emissions declined rapidly in the 1980s to its current low level (Urban & Nordensvard, 2018). Sweden is therefore a remarkable case of absolute decoupling of economic growth from carbon emissions. This evidence suggests that the Nordic countries, especially Denmark, Iceland and Sweden, can provide valuable lessons for national, regional and global low carbon energy transitions. A lot can be learned from these countries in terms of the technologies used, the infrastructure required, the financial instruments applied, the design and implementation of policy frameworks, the public and private governance of low carbon energy transitions, as well as how to achieve behavioural change and societal consensus for low carbon transitions. Other countries that have shown absolute declines in carbon dioxide emissions since the 1990s are Germany, the UK and France.

It has to be noted here that to go one step further and achieve carbon neutrality, low carbon transitions beyond the energy and heat sectors will have to be applied.

This is particularly difficult with regards to the industrial sector, which often relies on large-scale energy-intensive processes, such as processing of large amounts of raw materials and natural resources. In some countries this is happening increasingly, such as HYBRIT, the joint initiative of leading Swedish steel, mining and energy firms to produce fossil-free steel produced from hydrogen that is generated from renewable energy. Another tricky sector is the transport sector, where behavioural change is often needed as well as investments for infrastructure and financial incentives (such as subsidies for electric vehicles).

From traditional biomass to fossil fuels to low carbon energy

Traditional biomass

This section starts by examining traditional biomass as the first 'step' of an energy transition.

Traditional biomass includes **fuelwood** (often also referred to as firewood), **charcoal**, **dung** and **agricultural residues**. As mentioned in Chapter 1, 2.7 billion people worldwide rely on traditional biomass – such as fuelwood and dung – for basic needs such as cooking and heating (IEA, 2019). Traditional biomass combustion causes *indoor air pollution*, which can lead to negative health effects such as lung cancer, chronic respiratory diseases and adverse pregnancy outcomes. It is estimated that indoor air pollution from the combustion of solid fuels such as fuelwood causes 2 million deaths per year, of which about half are children (UNDP & WHO, 2009). This will be discussed in detail in Chapter 5. Another issue that is linked to traditional biomass use is the availability of fuelwood and its impact on deforestation. It is well established that fuelwood collection and charcoal production contributes to deforestation (Arnold et al., 2006).

While data on fuelwood production exist, it is difficult to estimate exactly how much fuelwood is being used by individuals, households and nations. Fuelwood consumption roughly equals fuelwood production, nevertheless most of the fuelwood does not enter official statistics and is not monetised, as it is often collected in woodlands free of charge. Global statistical databases, such as those of the FAO, account only for fuelwood, not for other forms of traditional biomass such as dung or agricultural residues. Despite the caveats, the data that are available indicate that fuelwood production per capita is decreasing with increasing income levels in all world regions (Van Ruijven et al., 2008; FAO, 2014). Nevertheless, absolute fuelwood production has increased over time in most regions. This is linked to growing populations that put pressure on scarce natural resources such as forest products, energy sources and food production. Exceptions are the countries of the Middle East that have abundant oil resources and little forest available (Van Ruijven et al., 2008; FAO, 2014).

Most of the technologies used for combustion traditional biomass are very simple. The simplest version is an open fire (sometimes referred to as a three stone fire). Improved cooking stoves have been introduced to make the combustion of traditional biomass more efficient. They use less fuel and reduce the amount of indoor air pollution.

Historically, observed energy transitions occurred from traditional biomass to fossil fuels. The next section will discuss fossil fuels.

Fossil fuels

Today about 80% of the world's primary energy supply comes from fossil fuel resources such as **oil, coal and natural gas** (IEA, 2019).

For **global energy production** between 1971 and 2017, the IEA estimates that the production of coal and natural gas has increased in the last few decades. The data also shows that global energy production has more than doubled between the 1970s and today (IEA, 2019). This has led to a sharp increase in global GHG emissions due to the heavy reliance on fossil fuels for energy production.

There has been a large increase in **global electricity generation** between 1990 and 2016. Particularly the electricity generation from coal and natural gas has increased rapidly in the last few decades. Coal is being used on a large scale for electricity generation in many emerging economies, including China, India and South Africa, as well as in Russia. Natural gas is being used on a large scale for electricity generation in several OECD countries such as the Netherlands and the UK. The figure also shows that global electricity generation has more than quadrupled between the 1970s and today (IEA, 2019). This has several causes, including population growth with more people using electricity, higher electrification rates giving more people access to electricity, a rise in electricity consumption due to an increase in global incomes, excessive electricity consumption by the richest segments in society, etc. This rapid rise in electricity generation has led to a sharp increase in global GHG emissions due to the heavy reliance on fossil fuels for electricity generation.

Coal, such as brown coal or lignite, is mainly used for heating, cooking and electricity generation. The energy efficiency of coal is lower than for natural gas and oil. Some coal is of low quality due to its high sulfur levels, such as those produced and predominantly consumed in China. Coal is mainly used for base-load electricity production. Coal is the most polluting of the fossil fuel energy sources, in terms of its contribution to global warming and climate change, as well as its contribution to local and transboundary air pollution as well as acid rain. The *emission factor* for coal ranges from 94,600 to 98,300 CO_2 kg/TJ as calculated in the IPCC guidelines on emission factors (IPCC, 2006). It has a higher emission factor than oil and natural gas. Coal is still being used by many poorer people as the preferred fuel for cooking and heating, even in upper-middle-income countries and emerging economies, such as China.

Besides **conventional coal-fired power plants** there are more efficient, super-critical coal power plants available such as the *integrated gasification combined cycle (IGCC)* power plants. Their efficiencies are much higher than the average efficiency for a conventional coal-fired power plant. In general, the efficiencies of coal-fired power plants are rather low compared to oil- and natural-gas-fired power plants. The newest IGCC plants are usually equipped with so-called *carbon capture and storage (CCS)* technology. Some people call these 'clean coal' technologies, although it needs to be kept in mind that no energy resource is completely clean and coal still remains a fossil fuel that is linked to a wide range of environmental problems.

Oil and oil-based products such as **petrol, diesel, kerosene and** *liquefied petroleum gas (LPG)* are mainly being used for transport. Petrol and diesel are usually used for vehicles while kerosene is mainly used for aviation, as it is much more energy-dense than other fuels. Oil is sometimes used for heating and for generating electricity, although its energy efficiency is lower than for natural gas. Oil is mainly used for base-load electricity production, although it can be used for peak-load production too. The emission factor for oil products ranges from 63,100 to 77,400 CO_2 kg/TJ as calculated in the IPCC guidelines on emission factors (IPCC, 2006). It has a higher emission factor than natural gas, but a lower emission factor than coal, which means that it contributes more to global warming than natural gas but less than coal.

Conventional oil-fired power plants are more efficient than coal-fired power plants, but less efficient and less flexible than natural-gas-fired power plants. The majority of today's electricity comes from coal or natural gas, whereas oil is mainly used for transport purposes and oil-based products such as kerosene or diesel are used for running private generators rather than large-scale electricity generation for the national grid.

Natural gas and natural-gas-based products, such as *compressed natural gas (CNG)*, are used for cooking, heating, electricity generation and transport. Natural gas is the preferred fuel for cooking and heating in many high-income countries. CNG is used for transport, such as for cars. Natural gas is also used for electricity generation; however, unlike coal it is often used for peak-load electricity generation. Natural gas power plants are extremely flexible: easy to start and shut down. They can therefore be adjusted to the specific electric load needs at peak times, such as in the mornings and evenings when there is a surge in demand as people wake up or when it gets dark in the evening. Natural gas has the highest energy efficiency of all fossil fuels and the lowest emission factor. The emission factor for natural gas ranges from 56,100 to 64,200 CO_2 kg/TJ as calculated according to the IPCC guidelines on emission factors (IPCC, 2006). This means that it contributes less to global warming than coal and oil. Natural gas is often considered a 'transition fuel' to shift from polluting fossil fuels, such as coal and oil, to cleaner, less polluting fuels. Nevertheless, natural gas is still a fossil fuel and contributes to global climate change and other environmental problems. A transition to low carbon energy requires a shift in infrastructure, politics, consumer behaviour and common practices that will go far beyond switching to natural gas.

Gas-fired power plants usually take the form of *combined cycle gas turbine (CCGT)* or *GT peak*. GT peak is a very flexible gas turbine that can be switched on and off within minutes depending on peak demand. As it is in gaseous form, it does not involve the combustion of a fuel and thereby makes it more flexible from a timing and ease-of-handling perspective.

Fossil fuels are heavily subsidised. Conservative data from the OECD suggests that about US$775 billion were spent on *fossil fuel subsidies* in 2012, with more than 80% of the subsidies being spent in developing countries, whereas the remaining share are fossil fuel subsidies in industrialised countries and in global production. Most of the fossil fuel subsidies were for oil-based products, followed by natural gas and coal (OECD, 2012). Fossil fuels subsidies distort markets and energy prices. Highly subsidised fossil fuels are mainly cheaper than renewable energy because of their artificially lowered prices. Without these subsidies, fossil fuel prices would be much higher and would make low carbon energy options more economically competitive. At the same time, renewable energy is subsidised in some countries through a feed-in tariff. We will discuss this in more detail in Chapter 9 when we will talk about the economics of energy and development.

Fossil fuels are based on finite, depletable resources that have formed over millions of years. Their depletion has severe environmental impacts, including the destruction of natural ecosystems and its impacts from coal mining and oil field exploration, as well as the harsh conditions for miners that often accompany fossil fuel depletion. For limiting global warming to a maximum of 2 degrees C, it is estimated that two thirds of the global fossil fuel resources need to stay in the ground. We will discuss the related environmental issues in detail in Chapter 6.

Low carbon energy

The IPCC states that energy-related GHG emissions from electricity generation, heat supply and transport account for about 70% of total GHG emissions (IPCC, 2007). Energy use has potentially significant climate impacts, which are assumed to exceed the impacts from other sources like land use and other industrial activities. So-called low carbon energy technologies, such as renewable energy technology including large hydropower, nuclear energy technology and CCS, are therefore key mechanisms to reduce carbon dioxide and other greenhouse gas emissions. Low carbon energy emits less GHGs than conventional fossil fuels such as coal, oil and natural gas. Some low carbon energy technologies, such as large hydropower and nuclear energy, have witnessed a recent revival due to climate change concerns. Nevertheless, this brings other adverse effects with it, such as concerns about health, safety and environmental impacts with regard to nuclear power.

Renewable energy comes from renewable natural resources, such as sunlight, wind, water, tides, geothermal heat and biomass. Unlike fossil fuels and nuclear energy, which are finite and depletable, these energy resources are renewable and non-depletable. Renewable energy has a large global potential. The World Energy Council estimates that the theoretical potential for solar energy is 370 petawatt hour per year (PWh/year), for primary biomass 315 PWh/year, for wind energy 96 PWh/year and for hydropower 41 PWh/year. Nevertheless, the technical and economic potential is lower due to variations in land availability and financial competition with fossil fuels (WEC, 2007). The most widely used and commercialised renewable energy technologies are wind turbines, solar PV panels and hydropower technology.

Globally, less than 20% of the total primary energy supply comes from non-fossil fuels. This is composed of about 5% nuclear energy, 2% hydropower, 10% biofuels and waste and 1% wind, solar and geothermal (IEA, 2018). The contribution of renewable energy to global electricity supply is, however, higher. Today renewable energy sources supply nearly 25% of global electricity (IEA, 2019), mainly from hydropower, but also from wind, solar and geothermal energy as well as from biomass. In some countries, renewable energy plays a much more important role, for example in the Nordic countries like Norway, Denmark, Sweden, Iceland and Finland.

Low carbon energy has had massive growth rates in the last few decades, both nuclear and renewable energy, particularly after the year 2000. As a comparison, the global gross electricity generation output from renewable energy was 2,364,468 GWh in 1990, 2,954,190 GWh in 2000 and 4,554,717 GWh (or 4,554.717 TWh or 4.554 PWh) in 2011 (IEA, 2018). The rate of wind and solar energy expansion has been particularly rapid in the last decade.

Low carbon energy includes the following energy technologies:

- *Wind energy*: Wind energy technology converts wind into electric power. It thereby converts kinetic energy into mechanical energy. Both *onshore* (on land) and *offshore* (in the sea) wind energy have increased rapidly since the 1990s. In 1996, only 6.1 gigawatt (GW) were installed and in comparison to 2017 when 432 GW were installed (WEC, 2017). The countries with the largest installed wind energy capacity are the USA, China, Germany, Spain, India and the UK (WEC, 2017). Wind turbines can range from small turbines, which are usually in the range of kilowatts (kW) and located onshore to large offshore turbines, which are in the range of several megawatts (MW) and located offshore. One modern large wind turbine can power up to 5000 households that have an average European energy demand, such as Enercon's E-126 turbine which has a generation capacity of 7 MW. Wind turbines can be connected to the central grid or used in mini-grids or as decentralised stand-alone systems.
- *Solar energy*: Solar energy technology converts solar radiation into electric power or heat. There are various forms of solar energy technology, including *solar photovoltaic (PV) panels* and modules, solar water heaters, solar thermal technology, solar lamps and solar cookers (used in some developing countries for cooking). Solar water heaters can often be found on rooftops of individual houses or buildings, whereas solar thermal technology, such as *concentrated solar power (CSP)*, is often used on a large-scale to replace fossil fuel power plants. Solar PVs can be found both on individual buildings and on a large scale to replace fossil fuel power plants. In 2017, the global installed capacity was 227 GW compared to 10 GW in 2007. The regions with the largest installed

solar PV capacity are Europe, particularly Germany and Italy, Asia, particularly China and Japan, and North America (WEC, 2017).

- *Hydropower*: Hydropower technology converts the energy of falling water into electricity or mechanical energy. Hydropower is the most widely used and commercialised renewable energy technology by far. In 2017, it supplied more than 16% of global electricity with an installed capacity of 1.21 TW (WEC, 2017). There is a distinction between *large hydropower* and *small hydropower*. Large hydropower refers to hydropower technology that has a generation capacity of more than 10 megawatt (MW) and often involves the construction of large dams and reservoirs. Dams have been the centre of much controversy in the 1990s and early 2000s due to their potential for severe environmental and social destruction; nevertheless, they are currently experiencing a renaissance due to the need for low carbon energy in the light of climate change. Small hydropower refers to hydropower technology that has a generation capacity of below 10 MW and is predominantly based on river run-off (ESHA, 2009). It is recognised as having low environmental and social impact compared to large hydropower schemes.

- *Nuclear energy*: Nuclear energy technology today uses nuclear fission to generate electricity and heat. At a global level, there were 390 GW installed in 2017 (IEA, 2019). The countries with the largest installed capacity of nuclear power are the United States, France, China, Japan (with very reduced capacity due to the Fukushima nuclear accident in 2011), Russia, South Korea, Canada and Ukraine. Other countries with considerable nuclear operating capacity are Germany (all nuclear power plants to be phased out by the end of 2022), the United Kingdom, Sweden, Spain, Belgium and Taiwan (WNA, 2019). While nuclear energy is a so-called low carbon energy source, producing less carbon and other GHG emissions than fossil fuels, it is fiercely debated due to severe health, safety and environmental implications. Two key controversies were exemplified in the nuclear disasters in Chernobyl, Ukraine (part of the former USSR), in 1986 and Fukushima, Japan, in 2011. Even today, more than 25 years after the nuclear accident in Chernobyl, the affected area – including its water, soil, flora and fauna – is heavily contaminated. Another problem with nuclear energy is that it depends on uranium, which is a finite resource.

- *Modern biomass*: Modern biomass, such as biofuels and biogas, is derived from organic material and is used for transport purposes (bioethanol and biodiesel), cooking (biogas) or electricity generation (biogas, waste-to-energy, wood chips, pellets, etc.). Biogas and waste-to-energy are usually derived from agricultural or livestock residues, such as poultry litter, which is then gasified and incinerated. This can take the form of large biomass power plants that feed electricity into the grid, for example in the UK and in Germany, or small-scale decentralised biogas production that is mainly used locally, for example in rural India. Biogas is used in several South Asian countries as a cleaner, more efficient and healthier cooking option than traditional biomass.

Biofuels are much more controversial; they include first-generation, second-generation and third-generation biofuels. *First-generation* or *conventional biofuels* are usually formed from edible biomass-based starch, sugar or vegetable oil. This means these fuels are usually based on food products, such as corn, wheat or other cereals, cassava or sugar beets that are used for making bioethanol. Brazil, for example, has had very successful bioethanol programmes in place for many decades, which has made it the global forerunner in bioethanol production with the world's largest vehicle fleet operating on bioethanol. As Brazil's bioethanol comes from sugar cane, it does not threaten food security (Dequech-Neto & Heiss, 2013). Soy, jatropha and palm oil are used for making biodiesel. *Second-generation biofuels* are biofuels that are not edible, but they are based on feedstock. This can include, for example, municipal waste. *Third-generation biofuels* or advanced/unconventional biofuels usually do not depend on food or feedstock products but can be derived from algae, cellulose and other forms of plant biomass, which makes it harder to extract fuel (Goldemberg & Lucon, 2009). Many countries are increasingly producing biofuels for transport. Unfortunately, it has been reported that this may negatively affect *food security*. The food security issue arises because land used for biofuel production cannot be used for food production; hence, there is a conflict between land for biofuels and land for food production. To make things worse, some biofuel crops, such as sweet potato or cassava, are also food crops (Rathmann et al., 2010). At the same time, there are allegations that some biofuel operations, including some operated by wealthy corporations, have evicted poor people from their lands in developing countries to gain access to the land to grow biofuels. Hence, some biofuel developments are reported to be associated with so-called land grabs (Neville & Dauvergne, 2012; Oxfam, 2012).

We will discuss low carbon energy technologies in more detail in Chapter 8; here, the key technologies are only briefly mentioned.

While the prices for nuclear energy remain high, the prices for renewable energy have decreased hugely since the 1980s. Most renewable energy technologies are now on par with the grid (so-called grid parity), meaning that the levelised cost of electricity (LCOE) is about the same as fossil-fuel-generated electricity. Figure 3.2 shows how the costs of solar PV have declined in recent years. While solar, wind and other renewable energy is also subsidised in some countries, such as through the *feed-in tariff* in many European countries, many countries do not have subsidies for renewable energy.

In recent years, wind and solar energy have become cost competitive with fossil fuels, including natural gas, and cheaper than nuclear energy.

The major barrier to a low carbon energy transition is therefore not of an economic nature but of a political nature. Fossil fuel economies are deeply engrained in today's lifestyles, policies and practices, which make a transition to a low carbon economy inherently political and complex (Lockwood, 2013).

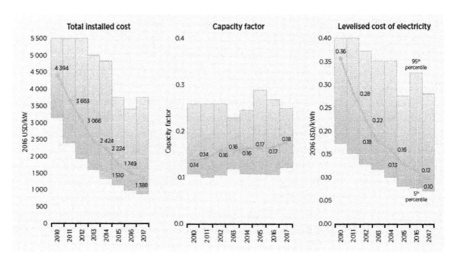

FIGURE 3.2   Prices for solar PV: installed costs, capacity factor and levelised costs of electricity (LCOE) between 2010 and 2017

Source: IRENA, 2018

Exercises

1   Which historic energy transitions have occurred in the past? How long were the time scales for these transitions? What was driving these transitions (see e.g. Bashmakov's three laws of energy transitions)? What does this potentially tell us about future energy transitions?

2   The levelised costs of electricity (LCOE) for renewable energy, particularly for solar PV, has reduced rapidly in recent years. What could this mean for low carbon energy transitions?

References

Arnold, J.E.M., Kohlin, G. & Persson, R. (2006) Woodfuels, livelihoods, and policy interventions: Changing perspectives. *World Development*, 34 (3), 596–611.

Bashmakov, I. (2007) Three laws of energy transitions. *Energy Policy*, 35 (7), 3583–3594.

Dequech-Neto, N. & Heiss, E. (2013) From outsider to world leader: Bioethanol in Brazil. In: Urban, F. & Nordensvard, J. (Eds.) *Low Carbon Development: Key Issues*, Earthscan, Routledge. pp. 284–296.

ESHA. (2009) *SHP in the World*. Small Hydro Power, European Small Hydropower Association (ESHA).

FAO. (2014) *FAO Statistics Database (FAOSTAT)*. Food and Agricultural Organization of the United Nations (FAO), Rome. Available from: http://faostat3.fao.org/home/E

Fouquet, R. & Pearson, P.J.G. (2012) Past and prospective energy transitions: Insights from history. *Energy Policy*, 50 (11/12), 1–7.

Fouquet, R. (2016) Historical energy transitions: Speed, prices and system transformation. *Energy Research & Social Science*, 22, 7–12.

Geels, F. (2002) Technological transitions as evolutionary reconfiguration processes: A multi-level perspective and a case-study. *Research Policy*, 31 (8/9), 1257–1274.

Geels, F. (2005) The dynamics of transitions in socio-technical systems: A multi-level analysis of the transition pathway from horse-drawn carriages to automobiles (1860–1930). *Technological Analysis and Strategic Management,* 17 (4), 445–476.

Goldemberg, J. & Lucon, O. (2009) *Energy, Environment and Development*. 2nd edition. Oxon, Earthscan, Routledge.

Grubler, A. (2012) Energy transitions research: Insights and cautionary tales. *Energy Policy*, 50 (11/12), 8–16.

IEA. (2019) *Statistics*. Paris, International Energy Agency (IEA), OECD/IEA. Available from: www.iea.org/statistics/

IPCC. (2006) *2006 IPCC Guidelines for National Greenhouse Gas Inventories. Volume 2: Energy*. Prepared by the National Greenhouse Gas Inventories Programme. Eggleston, H.S., Buendia, L., Miwa, K., Ngara, T. & Tanabe, K. (Eds.). Published: IGES, Japan.

IPCC. (2007) *Climate Change 2007: Mitigation. Contribution of Working Group III to the Fourth Assessment Report of the Intergovernmental Panel on Climate Change*. [Metz, B., O.R. Davidson, P.R. Bosch, R. Dave, L.A. Meyer (Eds.)] Cambridge University Press, Cambridge, United Kingdom and New York, NY, USA. Available from: www.ipcc.ch/ipccreports/ar4-wg3.htm

Lockwood, M. (2013) The political economy of low carbon development. In: Urban, F. & Nordensvard, J. (Eds.) *Low Carbon Development: Key Issues*, Earthscan, Routledge. pp. 25–37.

Lysen, E.H. (1996) *The Trias Energetica: Solar Energy Strategies for Developing Countries*. NOVEM, Utrecht.

Neville, K.J. & Dauvergne, P. (2012) Biofuels and the politics of mapmaking. *Political Geography*, 31 (5), 279–289.

OECD. (2012) *An OECD-Wide Inventory of Support to Fossil-Fuel Production or Use*. Organisation for Economic Co-operation and Development (OECD). Available from: www.oecd.org/site/tadffss/Fossil%20Fuels%20Inventory_Policy_Brief.pdf

Oxfam (2012) *Land Grabs*. Oxfam. Available from: www.oxfam.org.uk/get-involved/campaign-with-us/our-campaigns/grow/guide-to-land-grabs

Pearson, P.J.G. & Foxon, T.J. (2012) A low carbon industrial revolution? Insights and challenges from past technological and economic transformations. *Energy Policy*, 50, 117–127.

Rathmann, R., Szklo, A. & Schaeffer, R. (2010) Land use competition for production of food and liquid biofuels: An analysis of the arguments in the current debate. *Renewable Energy*, 35 (1), 14–22.

REN21. (2013) *Renewables 2013. Global Status Report*. Renewable Energy Policy Network for the 21st Century (REN21). Available from: www.ren21.net/Portals/0/documents/Resources/GSR/2013/GSR2013_lowres.pdf

Solomon, B.D. & Krishna, K. (2011) The coming sustainable energy transition: History, strategies and outlook. *Energy Policy*, 39 (11), 7422–7431

UNDP and WHO. (2009) *The Energy Access Situation in Developing Countries – A Review Focusing on Least Developed Countries and Sub-Saharan Africa*. United Nations Development Programme (UNDP) and the World Health Organization (WHO). Available from: http://content.undp.org/go/cms-service/stream/asset/?asset_id=2205620

Urban, F. & Nordensvard, J. (2018) Low carbon energy transitions in the Nordic countries: Evidence from the environmental Kuznets curve. *Energies*, 11 (9), 2209. doi: 10.3390/en11092209.

Urban, F. (2014) *Low Carbon Transitions for Developing Countries*. Oxon, Earthscan, Routledge.

Van Ruijven, B., Urban, F., Benders, R.M.J., Moll, H.C., Van der Sluijs, J., De Vries, B. & Van Vuuren, D.P. (2008) Modeling energy and development: An evaluation of models and concepts. *World Development*, 36 (12), 2801–2821.

WEC. (2007) *2007 Survey of Energy Resources*. World Energy Council (WEC). Available from: http://minihydro.rse-web.it/Documenti/WEC_2007%20Survey%20of%20Energy%20 Resources.pdf

WEC. (2017) *World Energy Resources Data*. World Energy Council (WEC). Available from: www.worldenergy.org/data/resources/

WNA. (2019) *World Nuclear Power Reactors & Uranium Requirements*. World Nuclear Association (WNA). Available from: www.world-nuclear.org/info/Facts-and-Figures/World-Nuclear-Power-Reactors-Archive/Reactor-Archive-March-2012/#.UgN1cY2kosQ

World Bank (2018) *Data: World Development Indicators*. The World Bank, Washington DC. Available from: http://data.worldbank.org/data-catalog/world-development-indicators

4

SECTORAL ENERGY NEEDS AND HOUSEHOLD ENERGY

The power sector

The Intergovernmental Panel on Climate Change (IPCC) estimates that about 70% of all greenhouse gas (GHG) emissions worldwide come from energy-related activities. This is mainly from fossil fuel combustion for heat supply, electricity generation and transport and includes carbon dioxide, methane and some traces of nitrous oxide (IPCC, 2007). It is estimated that the power sector alone accounts for about 45% of global emissions (IEA, 2019). The power sector and energy generation therefore play an important role in reducing GHG emissions. Key technologies such as low carbon energy technology, particularly renewable energy technology, as well as energy-efficient technology are therefore key to mitigating climate change.

The power sector is the electricity-producing sector. It generates power for use by individuals, households, firms and organisations, as well as for economic activities such as industry, services, agriculture and transport. In many high-income countries, the power sector is rather efficient, advanced technologies are being used and pollution control technologies are in place, such as end-of-pipe technologies like flue gas desulfurisation. Many power companies (utilities) are privately owned and accountable to shareholders, although some state-owned enterprises (SOEs) may exist too. In many middle-income countries, there is a mix of advanced and dated technologies, energy efficiencies can vary and pollution control technologies may or may not be used. Power companies tend to be SOEs, while fewer private utilities exist. This also means less accountability towards shareholders and closer links to state funding and subsidies. Some middle-income countries are struggling with dated and inefficient technology, state-owned utilities that are regularly underperforming due to being underfinanced, electricity prices that are below operating costs and a large group of poor customers that are unable to pay (Urban et al., 2007). Low-income countries are faced with even larger problems as there can be a prevalence of dated technologies, low energy efficiencies and limited pollution control technologies. Access to modern energy technology is therefore crucial. Low-income countries can be disproportionately affected by a lack of investment in the power sector and limited state finances. This is sometimes combined with a lack of fossil fuel resources making the country unstable with regards to energy security and excessively dependent on imports. Mitigating climate change in the power sector requires an increased investment in renewable energy technology and low carbon energy technology, such as wind, solar, hydropower and modern biomass.

The transport sector

The transport sector includes all forms of transport, including private and public vehicles, such as cars, buses, motorcycles, trucks, railway trains, trams, metros, ships and aircraft. Transportation also covers alternative, zero-carbon modes of transport, such as cycling and walking.

The transport sector is reported to account for about 20% of global energy use and almost 25% of energy-related carbon dioxide emissions (IEA, 2009; ITF, 2010). However, there is a degree of uncertainty associated with calculating the exact amount of emissions from this sector. Very certain is that the transport sector has grown rapidly, due to increasing car ownership across the world, particularly in emerging economies. As more people enter the middle class and have the purchasing power to buy a car, car ownership is rapidly increasing. With it, energy use and carbon dioxide emissions are increasing rapidly. Transport energy use and transport-related emissions are much higher in high-income countries than in middle- and low-income countries. Low-income countries have historically low energy use and emissions from the transport sector, although this is changing too. The International Energy Agency (IEA) estimates that global energy use and carbon dioxide emissions from transportation are likely to increase by 50% by 2030 and 80% by 2050. This is very clearly unsustainable in the light of climate change (IEA, 2009). The transport sector therefore plays a crucial role in mitigating GHG emissions. The IEA therefore suggests a three-fold strategy of (a) shifting to more sustainable transportation modes (such as shifting from private car to public transport, cycling or walking); (b) increasing efficiencies (such as improving the energy efficiency of combustion engines in cars); and (c) using alternative fuels (such as using electric cars rather than oil-based combustion cars). Particular emphasis is given to new technologies based on electricity, hydrogen and biofuels. This includes electric vehicles (EVs) such as hybrid vehicles, plug-in and battery-powered EVs and hydrogen-driven fuel cell vehicles (IEA, 2009). This comes, however, with the reservation that battery-powered electric vehicles, plug-ins and fuel cell vehicles are still in various stages of research and development (R&D) and are not fully commercialised yet. Advanced biofuel-powered vehicles are also important for achieving a low carbon energy transition in the transport sector. Nevertheless, biofuels can have several disadvantages and might negatively affect food production, as we discussed above;

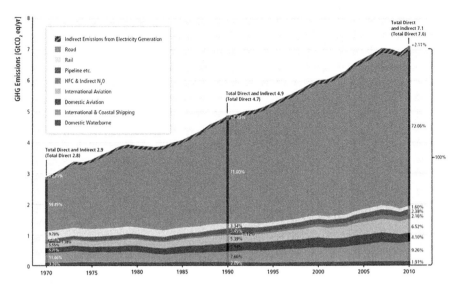

FIGURE 4.1 World transport energy use by mode between 1970 and 2010

Source: IPCC, 2018

therefore, it is important to acknowledge that first-generation biofuels (which are based on food products) should not be used for biofuels.

The graph in Figure 4.1 shows the increase in world transport energy use between 1970 and 2010. As the graph clearly shows, the majority of emissions originate from road transport. Concerning vehicle ownership in various countries, the USA is an outlier in comparison to other countries of similar GDP. The USA has far higher vehicle ownership per 1000 persons than other OECD countries. Increase in vehicle ownership in recent years is particularly rapid in emerging economies, such as in China.

The industrial sector

The industrial sector is the manufacturing or goods-producing sector and includes both heavy and light industry (it is often also called the secondary sector). Although there are no internationally agreed definitions, heavy industry usually refers to industry that requires high energy, material and chemical inputs and therefore has a larger environmental impact. It produces large products such as military equipment (e.g. tanks, fighter aircraft) and building equipment (e.g. cranes and bulldozers). It is often also more capital intensive. It includes, for example, the chemical industry, the oil and gas industry, the automotive industry and the aerospace and defence industry. Light industry, such as the home appliance, electronics and clothes industries, has a smaller environmental impact, is less capital intensive and requires less energy and fewer materials and chemicals.

According to estimates by the IPCC (2007) and World Resources Institute (WRI) (Herzog, 2009), the industrial sector accounts for almost 20% of global emissions. This goes beyond carbon dioxide emissions and includes methane, nitrous oxide, hydrofluorocarbons, perfluorocarbons and sulfur hexafluoride. Heavy industry, particularly the primary materials industry, is reported to account for about two thirds of the emissions. This is mainly the cement, iron and steel, chemical, petrochemical, paper and pulp, metal and mineral industries. However, there is a degree of uncertainty associated with calculating the exact amount of emissions from this sector. Nevertheless, it is clear that energy use and emissions from the industrial sector have increased rapidly in the last decade, particularly in China (WRI, 2005). This is partly due to the outsourcing and offshoring of industries from OECD countries to middle- and low-income countries such as China, hence energy use and emissions have increased in China, but the resources are consumed mainly in the EU and the USA. It is estimated that about a quarter of China's total carbon dioxide emissions are due to trade-related net export (Wang & Watson, 2007). Construction has also increased rapidly in China and other emerging economies due to a building boom. High-income countries used to have high energy use and high emissions from the industrial sector; however, this has declined over time as production is taking place elsewhere, such as in China. Low-income countries that are predominantly agrarian economies have low energy use and low emissions from industries.

The service sector

The service sector (often also called the tertiary sector) includes a wide range of services, such as consulting, education, financial services, healthcare, hospitality, legal services, management, media, real estate, retail, telecommunications and tourism. The service sector includes a wide range of 'white collar' employment. The energy use and GHG emissions from the services sector itself are relatively small. Differences in energy use exist between countries that are located in the (sub)tropics and those that are located in colder regions: Regions with a warmer climate have a higher energy demand for cooling such as air conditioning; regions with a colder climate have a higher energy demand for heating. Most of the energy use and the emissions are from energy generation for buildings, including offices and shops, as well as from transport use. The IPCC estimates that global GHG emissions from commercial and residential buildings together account for less than 10% of global emissions, although the figure is higher for carbon dioxide emissions (IPCC, 2007). Structural economic change from industries to services is therefore a useful strategy for a low carbon transition, but it has to be kept in mind that this often results in the outsourcing and offshoring of industry to less wealthy countries.

Cities

Cities are considered the motors of economic growth. A total of 85% of global gross domestic product (GDP) was generated in cities in 2015 (Gouldson et al., 2015).

Yet, cities are also the motors of environmental pollution and they might even be a threat to sustainable development. Urbanisation has led to environmental problems, such as high consumption of natural resources resulting in shortages, air/water/soil pollution, the emission of greenhouse gases contributing to climate change, reduction in green spaces and biodiversity, urban sprawl and excessive waste generation (Stern, 2007). Cities may be characterised by more unsustainable production and consumption patterns in comparison with rural areas. At the same time, cities are becoming increasingly vulnerable to the impacts of climate change (IPCC, 2007). Yet, there are an increasing number of cities that aim to be more sustainable and more environmentally friendly, such as Stockholm and Singapore, where the city government is aiming for more sustainable buildings, better public transport and a greener urban environment.

Cities are also major energy consumers and major emitters of greenhouse gases, mainly due to their large population, their high share of industries and services and their large transport sectors. In many cities, this increase in energy use and emissions is particularly driven by a rapidly growing transport demand and an expanding industrial sector. Low carbon strategies in the transport sector should therefore not only be limited to addressing individual car ownership and alternatives to petrol-driven vehicles, but should address city planning and urban development in its entirety. More charging stations for EVs need to be put in place in cities worldwide. Many cities around the world are plagued by sprawling urban settlements, long distances, strong dependency on private car ownership, limited use of public transportation, dominance of roads and hardly any infrastructure for cycling or walking. There is a need for cities that are less widespread, where attractive residential areas exist in city centres, where public transportation is a priority (underground, trains, trams, buses, boats) and cycling and pedestrian infrastructure exists, as well as a culture and societal value that encourages cycling and walking.

Gouldson et al. (2015) suggest that low carbon transitions in cities can be facilitated by prioritising policies and financial incentives that promote public transportation, building efficiency, increasing the share of renewable energy and having more efficient waste management systems in place.

The agricultural sector

The agricultural sector (often also called the primary sector) includes crop and livestock production. It is a key development sector as it contributes to food security and is a source of revenue for individuals, communities and countries. Many low- and middle-income countries have agrarian-based economies and depend on agriculture to a large extent. The agricultural sector is particularly important for the rural population as it also creates employment in rural areas. Nevertheless, in many poorer countries, work in the agricultural sector is often based on subsistence farming, unpaid work and informal labour. In high-income countries, the agricultural sector accounts for only about 1% of GDP and employs only a small number of people (World Bank, 2018).

At the same time, the agricultural sector is a key contributor to global climate change. This is partly due to carbon dioxide (CO_2) emissions, but also to methane (CH_4) and nitrous oxide emissions (N_2O). Agriculture is reported to contribute to about 15% of global anthropogenic GHG emissions (IPCC, 2007). Methane and nitrous oxide have far higher global warming potentials than does carbon dioxide, namely 28 times higher for methane and 265 times higher for nitrous oxide (IPCC, 2013). Agriculture-related emissions make up more than 50% of the world's total methane emissions and more than 80% of the world's total nitrous oxide emissions (Smith et al., 2008).

Land-use changes to create agricultural land also contribute substantially to GHG emissions, particularly if the land conversion happens from forest cover to agricultural land (deforestation). Deforestation reduces the amount of carbon stored in forest vegetation and soils – the carbon sink. Land-use changes refer to emissions from changes in land cover that reduce the amount of carbon stored in vegetation and soils (IPCC, 2000). Emissions from land-use changes are estimated to make up the largest share of the emissions from agriculture, followed by livestock and manure production.

Table 4.1 shows the sources of emissions from both direct and indirect drivers.

While agriculture plays an important role in global GHG emissions, the contribution of energy to these emissions is in fact rather low and is mainly limited to the emission of carbon dioxide. Low emissions transitions in the agricultural sector are possible (see e.g. Hiraldo, 2013), but this chapter focusses on energy issues only. Low carbon energy transitions in the agricultural sector are therefore similar to low carbon energy transitions in other sectors, namely achieved by introducing low carbon energy technology for transport and energy generation (particularly for irrigation) and increasing energy efficiency in agricultural processes.

TABLE 4.1 Greenhouse gas emissions from the agricultural sector, divided by direct and indirect drivers

Drivers of agricultural GHG emissions	Emission processes
N_2O from nitrogen fertilisers	Formed in soils
N_2O and CH_4 emissions from manure	N_2O is released when manure is applied to soils. CH_4 is produced due to anaerobic (oxygen-deprived) conditions.
CH_4 emissions from enteric fermentation	Produced as a by-product of animal indigestion
CO_2, CH_4 and N_2O from land use change	Emission due to land conversion, e.g. from pasture to crop land, deforestation, slash and burn agriculture
CH_4 from rice cultivation	Emissions from flooded rice paddies due to fermentation and decomposing organic matter
CO_2 from energy used for transport, manufacturing, agricultural activities	Emissions from the combustion of fossil fuels for farm machinery and vehicles, cooling of food products, production of fertilisers, etc.

Source: Amended from Hiraldo (2013: 177)

The household sector

The household sector includes the energy use of individuals, couples and families (households). Household energy demand is partly due to the need for energy for buildings, such as for lighting, heating, water heating, cooking, cooling and electronic appliances.

As mentioned before, climatic factors influence energy demand. Households in colder climates have a higher energy demand for heating, whereas households in warmer climates have a higher energy demand for cooling. Other household energy needs are for transport and the consumption of goods and services.

The IPCC estimates that global greenhouse gas emissions from commercial and residential buildings together account for less than 10% of global emissions, although the figure is higher for carbon dioxide emissions (IPCC, 2007). The energy demand and emissions from households that are linked to transport and the consumption of goods and services are more difficult to quantify as they are embedded in the figures of the transport, industrial, service and agricultural sectors.

Low carbon energy transitions in the household sector are possible (see, for example, the work of Urban [2009, 2014] on Beijing) and relatively easy and cost-effective to implement compared with other sectors. This is because energy use and emissions from the household sector are low compared with the energy use and emissions from the industrial sector, for example. Increasing the energy efficiency of buildings, such as introducing energy-efficient lighting and electronic devices, installing double- or triple-glazed windows and improved wall and loft insulation, can have a large impact and is common practice in EU member states. Many modern buildings also have rooftop solar PV or solar water heaters. The following section will address the issue of household energy for cooking and the transition from traditional biomass to fossil fuels to low carbon energy in that sector.

Household energy and household-based cooking

Household energy plays an important role in the development process. Household energy is directly linked to the Sustainable Development Goals (SDGs). For instance, ending poverty in all its forms everywhere (SDG 1) requires also ending energy poverty and enabling energy access for income-generation activities; eradicating hunger and achieving food security and improved nutrition (SDG 2), which only happen if cooking fuels are available for households to cook food; ensuring healthy lives and promoting well-being for all at all ages (SDG 3), which is only possible when fuels that are damaging health are reduced and when people have access to safe, clean and healthy energy options; achieving gender equality and empowering women and girls (SDG 4), which is linked to household energy use as women are mostly responsible for cooking and the energy provision for cooking (e.g. fuelwood collection); and protecting, restoring and promoting sustainable use of terrestrial ecosystems, sustainably managing forests, combating desertification, and halting and reversing land degradation, which is linked to sustainable management of energy resources (SDG 15). Most importantly, household energy links to SDG 7, the energy SDG that aims to ensure access to affordable, reliable, sustainable and modern energy for all (SDG 7).

Household energy differs between high-, middle- and low-income groups. It has been found that while low-income households use a lower number of energy services and have a lower number of end-uses, the primary energy technologies used in low-income households depend on a greater number of fuels and energy carriers, including dung, fuelwood, charcoal and liquefied petroleum gas (LPG). Middle-income households around the world mainly rely on electricity and natural gas, coal, LPG and kerosene. They use a higher number of energy services and have a higher number of end-uses. High-income households around the world consume much more energy than middle-income households but use the same technologies, fuels and energy carriers (Sovacool, 2011).

In many low- and middle-income countries, the majority of energy is used for cooking. In India, for example, approximately 80%–90% of all energy used in rural households is needed for cooking and heating water (Devi et al., 2009). This situation is similar in most sub-Saharan African countries. While many rural households spend more than three quarters of their energy demand on cooking, the situation is slightly different in urban areas, although they still spend a large amount of their energy demand on cooking. Even in the Chinese capital, Beijing, cooking accounts for almost 50% of the energy demand of urban household (Urban, 2014).

Cooking in low- and middle-income countries, and particularly in poorer households, is still predominantly done by using traditional biomass (IEA, 2010). Traditional biomass combustion causes indoor air pollution which triggers various adverse health effects and causes an estimated 2 million deaths per year, mostly of women and children (UNDP & WHO, 2009). A simple and cost-effective way of reducing the adverse effects of traditional biomass use from cooking is to introduce improved cooking stoves.

It is assumed that with increasing household income, households will switch from traditional biomass use to energy carriers that are cleaner, more efficient and easier to handle, such as from fuelwood to charcoal and then progressing on to more advanced energy carriers such as natural gas, biogas and electricity. This enables a transition from traditional biomass to modern fuels and energy carriers (Barnes & Floor, 1996; Holdren & Smith, 2000; Masera et al., 2000; Martins, 2005; Van Ruijven et al., 2008). This concept is called the energy ladder, which will be discussed in more detail later in this book. The energy ladder is linked to fuel switching, which is used to describe a process in which households switch from a fuel that is often inefficient and/or unhealthy to a fuel described as being modern and cleaner, such as from charcoal to LPG (Bloomfield & Yadoo, 2013). We will discuss this concept critically in Chapter 5.

Communities and energy use

This section focusses on the social practices and behaviours of individuals and communities that lead to carbon-intensive and less carbon-intensive lifestyles. It is widely acknowledged that human behaviour leads to environmental pressures, such as energy use, GHG emissions, air pollution, water use, land scarcity, etc. While behaviour is often individualised, there is a pattern that can be observed in communities. Communities are here defined as groups of people living in the same place or having similar characteristics, interests or attitudes. For example, the individual social practices of an average person in a low-income country such as Mali or Bhutan tend to be different to the individual social practices of an average person in a high-income country such as Australia or the UK. This relates to social and cultural practices and behaviour, which influence energy use and emissions. These individualised social practices and behaviour lead to aggregated, collective approaches to energy use and emissions in specific communities.

Social practices and behaviours need to be considered in the context in which they take place. They are shaped by norms, values, institutional arrangements, infrastructure and systems of governance. To understand energy use in specific communities, one needs to understand community practices that involve energy use and lead to emissions, as well as to be aware of the technologies, infrastructure and institutions available in these communities (Moloney et al., 2010). Typical energy use in communities is mainly related to household energy use, as well as transport energy use. Key relevant end-uses are therefore lighting, cooking, space heating, space cooling, and using appliances such as mobile phone chargers, computers, TVs and radios and kitchen appliances such as refrigerators, washing machines, dishwashers, microwaves, toasters, kettles and coffee machines. Obviously, there will be more end-uses and more appliances available to communities on higher incomes than on lower incomes. Transport energy use can range from the use of high carbon private vehicles, such as cars and motorcycles, to public transport, such as buses, trams and metros, to zero-carbon non-motorised modes of transport, such as walking and cycling. The individual behaviours and social practices determine the collective energy use and emissions within a specific community.

There has been lively debate for many years whether renewable energy is particularly useful for low-income communities in developing countries. The argument is often that these communities do not have access to modern energy (particularly electricity) and that they would benefit from access to renewable energy, often as a mini-grid or stand-alone option when connections to the grid are not an option. Bhattacharyya (2006) examines this argument and analyses whether renewable energy technology is an economically viable option for the poor and whether the poor are a major market for renewable energy technology. He concludes that the poor are not a major market for renewable energy technology, nor is it economically viable for them to use renewables, as long as prices for renewables are high and technologies are not mature yet. He argues that renewable energy technology has to compete against non-monetary energy options, such as fuelwood, and that renewable energy technology should therefore be given to the poor for free to make it a success. However, much has changed since Bhattacharyya's paper was published in 2006. It was published at a time when renewable energy technology was far less mature and far more expensive than it is today. Recent developments over the last decade or so have meant a rapid increase in the technological maturity of renewable energy technologies, such as wind, solar and hydropower, and renewable energy is now cost competitive with fossil fuels and cheaper than nuclear energy. In addition, there are the targets of the UN's Sustainable Energy for All (SE4ALL) initiative to achieve modern energy access with renewable energy for those living in energy poverty, which resonate in the energy target of the SDGs (SDG 7). Investments in large-scale energy infrastructure, such as large hydropower dams, have also increased electrification rates in some countries, such as in Asia.

Yet, there is a moral argument embedded in the thinking of Bhattacharyya (2006). The question is whether it is morally acceptable to ask the poor in developing countries to be content with renewable energy, which, after all, may have challenges such as fluctuating energy supply and issues with energy storage, whereas the wealthy in developed countries are guaranteed modern energy, which is predominantly based on fossil fuels (e.g. grid-based electricity, petrol for transportation).

Exercises

1   Which sectors contribute most to the emission of greenhouse gases and thereby to climate change? In which ways?
2   How can transformation of these sectors contribute to more sustainable development?
3   What are the challenges that are being faced in these sectors today and in the future?
4   Choose two countries of your choice, one high-income country and one low/middle-income country (e.g. USA and Cuba). Go to the IEA statistics website at www.iea.org/statistics. Check out the energy consumption for the country of your choice and click on share of total final energy consumption by sector. Which sectors consume the highest amount of energy and why? How do these two countries differ and why? Please note the IEA only breaks down transport, industry and other.

References

Barnes, D.F. & Floor, W.M. (1996) Rural energy in developing countries: A challenge for economic development. *Annual Review of Energy and the Environment*, 21 (1), 497–530.

Bhattacharyya, S.C. (2006) Renewable energies and the poor: Niche or Nexus? *Energy Policy*, 34 (6), 659–663.

Bloomfield, E. & Yadoo, A. (2013) Low carbon energy and energy access in developing countries. In: Urban, F. & Nordensvärd, J. (Eds.) *Low Carbon Development: Key Issues*, Earthscan, Routledge. pp. 132–150.

Devi, R., Singh, V., Dahiya, R.P. & Kumar, A. (2009) Energy consumption pattern of a decentralized community in northern Haryana. *Renewable and Sustainable Energy Reviews*, 13 (1), 194–200.

Gouldson, A., Colenbrander, S., Sudmant, A., McAnulla, F., Karr, N., Sakai, P., Halla, S., Papargyropoulou, E. & Kuylenstiern, J. (2015) Exploring the economic case for climate action in cities. *Global Environmental Change*, 35(11), 93–105.

Herzog, T. (2009) *World Greenhouse Gas Emissions in 2005.* World Resources Institute (WRI). WRI Working Paper. Available from: www.wri.org/publication/world-greenhousegas-emissions-2005

Hiraldo, R. (2013) Agriculture and low carbon development. In: Urban, F. & Nordensvärd, J. (Eds.) *Low Carbon Development: Key Issues*, Earthscan, Routledge. pp. 176–187.

Holdren, J.P. & Smith, K.R. (2000) Energy, the environment and health. In: *World Energy Assessment, Energy and the Challenge of Sustainability*. United Nations Development Programme (UNDP), United Nations Department of Economic and Social Affairs (UNDESA) and World Energy Council. pp. 62–110.

IEA. (2009) *Transport, Energy and CO2*. Paris, International Energy Agency (IEA), OECD/IEA. Available from: www.iea.org/publications/freepublications/publication/transport2009.pdf

IEA. (2010) World Energy Outlook 2010. *Energy Poverty: How to Make Modern Energy Access Universal?* Paris, International Energy Agency (IEA), OECD/IEA. Available from: www.worldenergyoutlook.org/media/weowebsite/2010/weo2010_poverty.pdf

IEA. (2019) *Statistics.* Available from: https://www.iea.org/statistics/

IPCC. (2000) *SRES Special Emission Scenarios*. Special Report of the Intergovernmental Panel on Climate Change (IPCC). [Nakicenovic, N. & Swart, R. (Eds.)], Cambridge, Cambridge University Press.

IPCC. (2007). *Climate Change 2007: Mitigation*. Contribution of Working Group III to the Fourth Assessment Report of the Intergovernmental Panel on Climate Change. [B. Metz, O.R. Davidson, P.R. Bosch, R. Dave, L.A. Meyer (Eds.)], Cambridge, UK and New York, NY, USA. Available from: www.ipcc.ch/ipccreports/ ar4-wg3.htm

IPCC. (2013) *Climate Change 2013: The Physical Science Basis*. Contribution of Working Group I to the Fifth Assessment Report of the Intergovernmental Panel on Climate Change [Stocker, T.F., D. Qin, G., K. Plattner, M. Tignor, S.K. Allen, J. Boschung, A. Nauels, Y. Xia, V. Bex and P.M. Midgley Geels, F. (2012) A socio-technical analysis of low-carbon transitions: Introducing the multi-level perspective into transport studies. *Journal of Transport Geography*, 24 (9), 471–482].

IPCC. (2018) *Graph on GHG Emissions from Transport*. Available from: www.ipcc.ch/site/assets/uploads/2018/02/01_figure_8.1.png

ITF. (2010) *Reducing Transport Greenhouse Gas Emissions. Trends and Data 2010*. OECD/ITF International Transport Forum, Paris.

Martins, J. (2005) The impact of the use of energy sources on the quality of life of poor communities. *Social Indicators Research*, 72 (3), 373–402.

Masera, O.R., Saatkamp, B.D. & Kammen, D.M. (2000) From linear fuel switching to multiple cooking strategies: A critique and alternative to the Energy Ladder Model. *World Development*, 28 (12), 2083–2103.

Moloney, S., Horne, R.E. & Fien, J. (2010) Transitioning to low carbon communities – from behaviour change to systemic change: Lessons from Australia. *Energy Policy*, 38 (12), 7614–7623.

Smith, P., Martino, D., Cai, Z., Gwary, D., Janzen, H., Kumar, P., McCarl, B., Ogle, S., O'Mara, F., Rice, C., Scholes, B., Sirotenko, O., Howden, M., McAllister, T., Pan, G., Romanenkov, V., Schneider, U., Towprayoon, S., Wattenbach, M. & Smith, J. (2008) Greenhouse gas mitigation in agriculture. *Philosophical Transactions of the Royal Society B*, 363, 789–813.

Sovacool, B.K. (2011) Conceptualizing urban household energy use: Climbing the 'Energy Services Ladder'. *Energy Policy*, 39 (3), 1659–1668.

Stern, N.H. (2007) *The Economics of Climate Change: The Stern Review*. Cambridge, UK, Cambridge University Press.

UNDP and WHO. (2009) *The Energy Access Situation in Developing Countries – A Review Focusing on Least Developed Countries and Sub-Saharan Africa*. United Nations Development Programme (UNDP) and the World Health Organization (WHO). Available from: www.undp.org/content/dam/undp/library/Environment%20and%20Energy/Sustainable%20Energy/energy-access-situation-in-developing-countries.pdf

Urban, F. (2009) Climate change mitigation revisited: Low-carbon energy transitions for China and India. *Development Policy Review Journal*, 27 (6), 693–715.

Urban, F. (2014) *Low Carbon Transitions for Developing Countries*. Oxon, Earthscan, Routledge.

Urban, F., Benders, R.M.J. & Moll, H.C. (2007) Modelling energy systems for developing countries. *Energy Policy*, 35 (6), 3473–3482.

Van Ruijven, B., Urban, F., Benders, R.M.J., Moll, H.C., Van der Sluijs, J., De Vries, B. & Van Vuuren, D.P. (2008) Modeling energy and development: An evaluation of models and concepts. *World Development*, 36 (12), 2801–2821.

Wang, T. & Watson, J. (2007) *Who Owns China's Carbon Emissions?* Tyndall Centre for Climate Change Research, Brighton. Briefing Note No. 23.

World Bank (2018) *Data: World Development Indicators*. The World Bank, Washington DC. Available from: http://data.worldbank.org/data-catalog/world-development-indicators

WRI. (2005) Cumulative emissions. In: *Navigating the Numbers: Greenhouse Gas Data and International Climate Policy*. World Resources Institute (WRI). pp. 31–33.

5

CONCEPTS OF ENERGY AND DEVELOPMENT

The energy ladder

The energy ladder is a stylised concept which assumes that with increasing incomes, households will be able to switch from traditional biomass use to energy carriers that are cleaner, more efficient and easier to handle, such as charcoal, and then progress on to modern energy carriers such as natural gas, biogas and electricity (Barnes & Floor, 1996; Holdren & Smith, 2000; Masera et al., 2000; Van Ruijven et al., 2008). Figure 5.1 shows how stove energy efficiency and stove capital cost increase with increasing affluence during the transition from traditional biomass, such as dung, agricultural residues, wood and charcoal, to improved cooking stoves, kerosene stoves, liquefied petroleum gas (LPG) stoves and finally the electric hot-plate. The figure indicates the relationship between per capita final energy consumption and income in developing countries, measured as the percentage of the population living on less than US$2 per day.

Organisation for Economic Co-operation and Development (OECD) countries are usually at the top of the energy ladder and consume mainly modern energy, such as electricity and natural gas. This is different in non-OECD countries. In real life, fuel switching as suggested by the concept of the energy ladder is far less discrete, and often various energy carriers are used alongside each other, such as electricity for lighting and charcoal for cooking (Masera et al., 2000). This is called fuel stacking. For example, it was found that in India several fuels are being used alongside each other in all income groups. Traditional biofuels are used predominantly, alongside charcoal, oil products such as LPG and kerosene, biogas, solar energy and electricity, were available (Urban et al., 2009). Once access to modern energy is provided, fuel switching may occur. While the poorer income groups tend to cook predominantly with traditional biofuels, the wealthier income groups

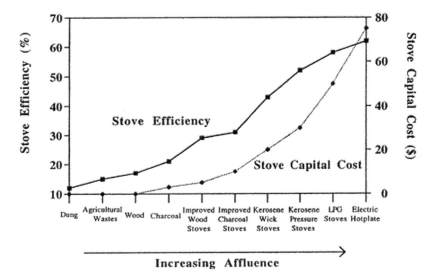

FIGURE 5.1 The energy ladder concept: the increase of stove efficiency and capital cost from traditional biomass to modern energy options

Source: Masera et al., 2000: 2084

tend to use cleaner and more efficient fuels such as LPG or biogas more frequently (Bhattacharyya, 2006a, 2006b).

There is evidence that several fuels are being used in real life at the same time by the same household, meaning that the concepts of the energy ladder and fuel switching have only been observed to some extent in reality. While it is often assumed that the rural poor will be those that are stuck at the bottom of the energy ladder, there is new evidence that suggests that the situation may not be that different for the rural rich. Mirza and Kemp (2011) suggest that the urban poor often have better access to more energy-efficient and high-quality energy than do the rural rich. The problem of the urban poor is that they cannot afford to pay for these high-quality fuels. The problem of the rural rich is that energy-efficient, high-quality fuels are often difficult to access and may involve consider-able time and transaction costs (Mirza & Kemp, 2011). One problem, as with the conceptualisation and the real-life evidence from historic energy transitions, is that the concept of the energy ladder is based on the experience of today's developed countries. Households in these countries have climbed the energy ladder over cen-turies, starting from the use of traditional biofuels and ending with using modern fuels, such as electricity and natural gas. However, the same linear transition cannot be observed in today's developing countries. This is due to a number of factors, including the availability of modern fuels alongside traditional fuels and issues of access, costs and technologies. Nevertheless, there is a clear link between access to electricity and income, as well as prevalence of traditional biomass and income, as discussed in Chapters 1–4. Despite the criticism, the concept of the energy ladder remains a valid theory based on historical evidence from developed countries and some observations in developing countries. In practice, however, it is less linear and is fuzzier than the stylised concept suggests.

Fuel switching

The energy ladder is linked to fuel switching. Fuel switching is used to describe a similar process where households switch from a fuel that is often inefficient and/ or unhealthy to a fuel that is cleaner and more modern, such as from charcoal to LPG or natural gas. Fuel switching essentially requires the displacement of one fuel by another and usually involves a transition from solid fuels to modern fuels such as liquid fuels and/or electricity. Heltberg (2004) suggests that fuel switching as conceptualised in the energy ladder involves three phases. The first phase is determined by a universal reliance on traditional biomass, meaning that households predominantly use fuelwood for cooking and other energy activities. The second phase requires fuel switching of households to enable an energy transition from traditional biomass to cleaner and more efficient fuels such as coal, charcoal and kerosene. This second phase is driven by higher incomes, urbanisation and biomass scarcity. The third phase is characterised by households switching to modern fuels such as natural gas, electricity, LPG or biogas for cooking. This is primarily driven by growing incomes in relation to fuel prices. In real life, fuel switching and the energy ladder are far less discrete and linear. Fuel switching is contested by leading academics (e.g. Masera et al., 2000; Heltberg, 2004). Often various energy carriers are used alongside each other, such as electricity for lighting and charcoal for cooking (hence fuel stacking occurs). Multiple cooking strategies have also been observed (Masera et al., 2000). Heltberg (2004) makes the point that one therefore needs to differentiate between fuel choice and switching. For example, in Guatemala, LPG is used by about 25% of urban households and about 15% of rural households, while fuelwood is predominantly used, sometimes by the same households simultaneously. In South Africa, about 35% of rural households used both fuelwood and kerosene for cooking (Heltberg, 2004). This means that rather than 'climbing up the energy ladder' neatly from one step to the next, households in fact use a range of multiple fuels at the same time, both more efficient and less efficient ones, as well as cheaper and more costly ones. Masera et al. (2000) call this fuel stacking as opposed to fuel switching. There are three options for fuel switching: (a) no fuel switching happens; (b) partial fuel switching happens; or (c) full fuel switching happens. Fuel switching is advanced in urban areas where full fuel switching often happens with increased household income. In rural areas, fuel switching is rather rudimentary, as often only partial fuel switching or no fuel switching at all happens. The key factors that determine fuel choice and fuel switching are households' budgets, preferences and needs. More specifically, household incomes and expenditures matter. For example, the disposable income that households have available and its relationship to fuel prices is a key factor in influencing fuel choice (e.g. determining whether someone can afford to buy LPG and an LPG cooker or whether they cannot and is therefore collecting fuelwood which is available free of charge). A second key factor is obviously fuel prices. The higher the fuel prices the more pressure they pose for households who are struggling to make ends meet. Culture also plays a role. In some countries, cooking with fuelwood is cultural and traditionally ingrained, as these practices have been used for hundreds of years or more, while a switch to cleaner energy sources may also require a change in cooking practices.

Another deciding factor is urbanisation: the more urbanised households are, the more they are likely to switch to modern fuels. This is linked to accessibility of energy markets, services and infrastructure. This leads to a fourth decisive factor, namely the link between access to infrastructure, markets and services particularly in relation to access to electricity. There is an observed link between access to electricity and a decline in the use of fuelwood in urban areas. For rural areas, the correlation is less distinct. This is an interesting phenomenon, as electricity is rarely used for cooking and mainly for lighting and powering consumer appliances (Heltberg, 2004). An interesting finding is that education also plays a role in fuel choice. Heltberg (2004) found that there is a positive correlation between a higher level of education and a higher level of modern fuel use. This could be explained by several factors: For example, individuals with higher education have better jobs and higher incomes; or they are more educated about the negative impacts of traditional fuelwood use, such as indoor air pollution, or likewise about the positive effects of electricity access, such as the opportunity to study or work after dark; or due to their more demanding jobs they may not have the time available to collect fuelwood. This is a complex issue. In Brazil, it was found that only the education of the female spouse mattered concerning fuel switching, whereas the education of the male household head was insignificant (Heltberg, 2004). Heltberg (2004) finds that

> (1) modern fuel use relates positively to per capita expenditures; solid fuels are negatively related to expenditures. This holds in all of the countries studied and in both urban and rural areas;
>
> (2) modern fuel use is positively correlated with electrification of the household; usage of solid fuels declines in response to electrification;
>
> (3) having tap water inside the house is also associated with fuel switching in most instances;
>
> (4) larger households tend to use a greater number of fuels, both solid and non-solid;
>
> (5) increasing levels of education are associated with a higher probability of using modern fuels and a lower incidence of solid fuel use.
>
> *Heltberg (2004: 885)*

For policy and practice this means that energy interventions for increasing modern energy access should primarily be aimed at households that have the means to purchase modern fuels and modern energy equipment, that have high expenditure on solid fuels and that already have access to modern energy infrastructure, for example urban residents who have access to electricity. Energy interventions for households without access to modern energy infrastructure, which depend on cheap or free traditional biomass and do not have the means to purchase modern energy and equipment, may have limited success. It may be more effective to target these households through other interventions such as improved cooking stoves and better ventilation in kitchens to decrease indoor air pollution (Heltberg, 2004).

Concepts of economic growth and energy-related pollution control

The environmental Kuznets curve

In this section, we discuss the environmental Kuznets curve (EKC), an important concept for energy and development, which is relevant to understanding the process of development from an energy-related pollution perspective. The EKC is also important for understanding energy transitions, such as low carbon energy transitions, and for providing insights into how they relate to environmental pollution throughout the development process. As we discussed in Chapter 2, the economist Simon Kuznets developed the concept of the Kuznets curve showing that the development of an economy follows an inverted U-shaped development, where income inequality first increases with increasing gross domestic product (GDP) levels and, after a threshold, the income inequality declines with further increasing GDP levels (Kuznets, 1955; Goldemberg & Lucon, 2009). The Kuznets curve is highly debated in academic literature, with scholars affirming, rejecting and fiercely

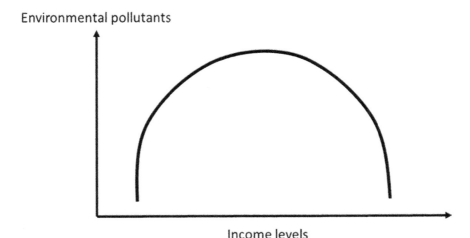

FIGURE 5.2 Stylised concept of the environmental Kuznets curve (EKC)

Source: Urban and Nordensvärd, 2018

discussing the concept and the phenomenon of increasing and decreasing income distribution with the development of a national economy (see, for example, Saith, 1983; Glomm, 1997; Deininger & Squire, 1998; Fields, 2001). Similar developments can be seen in the environmental field. The EKC is a contentious concept that correlates environmental pollution and economic development. The EKC has the shape of an inverted U-curve, similar to the income inequality curve described by Kuznets in the 1950s (Kuznets, 1955; Van Ruijven et al., 2008). The hypothesis of the EKC is that environmental pollution is at a low level when countries have very low income and development levels, then it increases and peaks when mid-levels of incomes and development are reached, and pollution levels decrease again when income and development levels increase (Van Ruijven et al., 2008). This is based on the assumption that pollution levels will increase when developing countries industrialise, but will decrease again when countries become more prosperous and can afford to invest in pollution control technologies. A stylised example of the EKC is shown in Figure 5.2. The EKC is closely related to the decoupling of economic growth from energy use and emissions, which is fiercely debated (Urban & Nordensvärd, 2013, 2018).

Some of the controversy surrounding the EKC is related to whether the EKC can be observed in practice and how it can be measured; various approaches to units and measurements exist. Nevertheless, the EKC has been historically observed in many countries for pollutants such as sulfur dioxide (SO_2), where end-of-pipe technologies are relatively inexpensive and easy to add. However, the EKC has not been observed for carbon emissions on a global level (Van Ruijven et al., 2008). It has rather been observed in reality that carbon emissions increase with increasing levels of income and development, and then either continue to increase, or level off or there is a moderate decline, but not a full U-curve similar to the EKC concept. The EKC basically assumes that pollution levels are low in low-income countries, increasing and peaking in middle-income countries and then decreasing in high-income countries. In real life, this assumption is skewed by the offshoring and outsourcing of polluting industries from high-income countries to low- and middle-income countries. Historically, countries such as Taiwan were the manufacturing hub for Western developed countries in the 1980s and early 1990s, then China in the late 1990s and 2000s and after 2010, manufacturing (e.g. for textiles) moved increasingly to lower-middle-income countries such as India and low-income countries, such as Cambodia and Bangladesh. Not only are large shares of the Western countries' clothes, textiles, electronics and other consumer goods manufactured in low- and middle-income countries, but in turn Western firms also 'export' their greenhouse gas (GHG) emissions and environmental pollution to these countries. This is one way in which globalisation defies the EKC. Essentially, achieving the EKC requires a decoupling of economic growth from emissions. For mitigating climate change in the long run, a decrease in emissions is needed; however, this proves to be difficult as the experience of the EKC and absolute decoupling of economic growth from emissions suggest (Urban & Nordensvärd, 2013).

Technological leapfrogging

Technological leapfrogging is a concept that is related to the EKC. It assumes that countries can learn from the mistakes of others and 'leap' over the most polluting phases of development, such as the coal era during the Industrial Revolution, which Europe and North America had to go through. Low- and middle-income countries may be able to avoid such periods due to cleaner and more advanced low carbon technologies. In visual terms, this would look like the graph in Figure 5.3. The grey arrow shows how a country would increase income levels and increase its development status while at the same time leaping over the most environmentally polluting phase. This can happen through improved access to modern, less polluting technologies. Goldemberg and Lucon (2009) mention several cases in which technological leapfrogging has happened in relation to energy and development:

- In off-grid communities in sub-Saharan Africa, for example in Ghana, solar photovoltaic (PV) panels have been used as an alternative to fuelwood and diesel generators. Byrne et al. (2014) also found cases of technological leapfrogging in Kenya concerning solar PV and solar lamps.
- In China, energy-efficient domestic appliances have been taken up on a large scale, for example, energy-saving light bulbs.
- In Brazil, flexible-fuel vehicles have been introduced on a large scale, which enables the widespread use of bioethanol-driven vehicles.

Technological leapfrogging is, however, subject to the transfer of modern technologies from high-income countries to low- and middle-income countries and

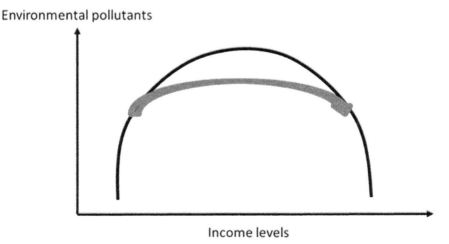

FIGURE 5.3   Stylised example of technological leapfrogging

Source: Urban and Nordensvärd, 2018

subject to having adequate skills, knowledge and expertise in the recipient countries to use these technologies adequately in the long term. This is essentially not only a question about technology transfer, especially hardware, but it is a question about building up indigenous innovative capacity, including skills, knowledge and expertise to absorb, deploy and develop modern energy technology. Long-term technological leapfrogging is also dependent on adequate investments and financing options.

The United Nations Framework Convention on Climate Change (UNFCCC), the Kyoto Protocol and the Paris Agreement embed the issue of technology transfer and technological leapfrogging in the international climate change agenda. For example, this is done by means of the Clean Development Mechanism (CDM) under the Kyoto Protocol. The use of the CDM can in principle enable low- and middle-income countries (non-Annex I countries) to receive modern technology for sustainable development from industrialised countries (Annex I countries), while industrialised countries, in return, receive certified emission reduction credits (CERs) to offset their emission reduction obligations. This is said to encourage sustainable development in non-Annex I countries. Nevertheless, many non-Annex I countries and scholars have criticised the CDM because it does not meet its goals and only generates a transfer of hardware or financial resources rather than contributing to sustainable development (Karp & Liu, 2000; Sutter & Parreño, 2007; Ockwell et al., 2008; Boyd et al., 2009; Nussbaumer, 2009; Drupp, 2011; Srinkanth & Lloyd, 2011). This is exemplified by the fact that the large majority of CDM projects go to emerging economies, mostly China, followed by India, whereas only a tiny share is allocated to low-income countries. The whole of sub-Saharan Africa (excluding South Africa) receives less than 1% of CDM projects. In general, countries that have a stable investment climate, a sizeable economy and access to modern low carbon technology, such as China, India, Brazil, Mexico and South Korea, have a much higher share of CDM projects than have other countries.

Byrne et al. (2014) argue for smaller-scale, pro-poor low carbon technology transfer that aims to build up the indigenous innovative capacity of the recipient country as opposed to the large-scale, top-down initiatives of the CDM that mainly generate access to hardware and financial transactions. This implies that technological leapfrogging is dependent not only on the access to advanced, less polluting technologies, but also needs to build up local capacities and be of value to people in low- and middle-income countries who can use these technologies on a daily basis.

Decoupling economic growth from energy use and emissions

This section discusses another crucial issue for energy and development, namely the decoupling of economic growth from energy use and greenhouse gas emissions. It is often argued that decoupling growth from energy use and emissions is important for mitigating climate change while still achieving development. This is important for the transition from traditional biomass to modern fuels as well as for the transition from high carbon development to low carbon development. Decoupling of economic growth from energy use and/or emissions requires that at some point in time the growth rate for energy use and/or emissions is lower than the GDP growth rate. Absolute decoupling requires an absolute cut in energy use and/or emissions, which is difficult to achieve. Only a few countries have achieved an absolute decoupling of economic growth from carbon emissions over long periods of time. They include Sweden, Denmark, Iceland, Germany, the UK and France. In Sweden, the carbon emissions today (2019) are lower than they were in the 1960s. Denmark has even achieved a decoupling of economic growth from energy use, which is a rare development (Urban & Nordensvärd, 2018).

What has been observed for most countries and certainly on a global level, however, is a clear correlation between economic growth and GHG emissions. Both have strongly increased over the last 100 years worldwide (IPCC, 2007), which shows that absolute decoupling has not happened on a global level, even though there are exceptions, such as those just discussed (Urban, 2010; Urban & Nordensvärd, 2013). Relative decoupling means that more economic activity is possible with lower energy use and/or lower emissions. This is measured, for example, in carbon intensity or energy intensity, which is the amount of carbon emissions or energy used per unit of GDP. Relative decoupling in terms of carbon and energy intensity has been observed worldwide since the 1960s. Other examples are the OECD countries as an aggregate and also China and India, for example, which have rapidly decreasing carbon and energy intensities (Van Ruijven et al., 2008; Urban, 2010; Urban & Nordensvärd, 2013). Yet, this has not resulted in an absolute decline in carbon emissions; instead, they have rapidly increased in recent decades worldwide (IEA, 2019). The last few years have seen a stagnation of total CO_2 emissions in China (IEA, 2019), which is a very positive development as it might mean that China's emissions may have peaked and that they may be declining from now on. It remains to be seen what happens in future years.

The issue of decoupling is fiercely debated. Some advocate that decoupling growth from energy use and emissions is only possible to some extent due to physical limits (Ockwell et al., 2008) and that instead the structure of market economies has to be changed to achieve deep cuts in emissions (Jackson, 2009). Jackson of the UK Sustainable Development Commission even talks about the 'myth' of decoupling (2009: 8, 46 onwards). Others advocate that decoupling is possible when low carbon and energy-efficient technology is used (Barrett et al., 2008). Decoupling economic growth from carbon emissions is a complicated issue; however, decoupling economic growth from other GHG emissions, such as from methane and nitrous oxide, is also very difficult and the evidence is under-researched and poorly understood. In fact, most of the decoupling debates do not take into account the emissions from deforestation and land-use changes due to these pitfalls (Urban, 2010). While there are discussions about the limits of decoupling growth from carbon emissions, many case studies argue that low carbon growth is possible, for example in China (IEA, 2007), India (World Bank, 2008), South Africa (Government of South Africa, 2008) and Mexico (Project Catalyst, 2008; Urban, 2010). Data analysis from the Nordic countries, particularly Sweden and Denmark, also suggests that economic growth can be decoupled from carbon emissions over longer periods of time (Urban and Nordensvärd, 2018).

Figures 5.4 and 5.5 indicate how relative decoupling of economic growth from carbon emissions has been achieved worldwide and in the EU, whereas an absolute decoupling has not been observed worldwide and in the OECD over longer periods

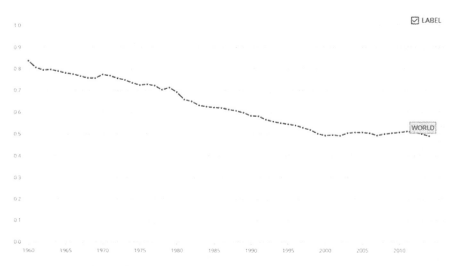

FIGURE 5.4 Relative decoupling has been achieved since the 1960s world-wide, measured as carbon intensity (kg CO_2 per unit of GDP)

Source: Data from World Bank, 2019

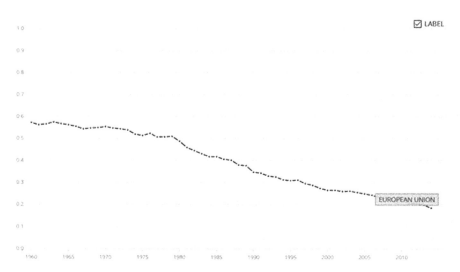

FIGURE 5.5 Relative decoupling has been achieved since the 1960s for EU countries, measured as carbon intensity (kg CO_2 per unit of GDP)

Source: Data from World Bank, 2019

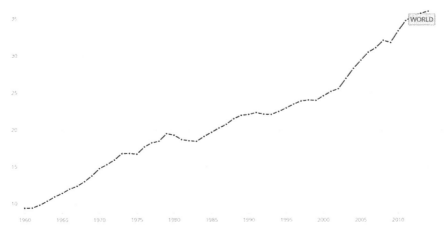

FIGURE 5.6 Absolute decoupling has been not been achieved since the 1960s world-wide, measured in Mt CO_2 emissions

Source: Data from World Bank, 2019

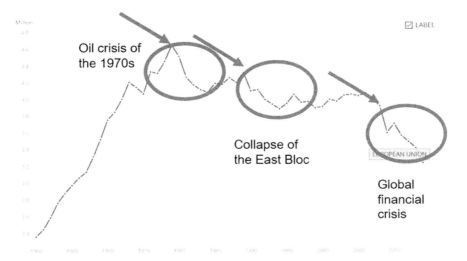

FIGURE 5.7 Absolute decoupling has been achieved since the 1960s in the EU (as an aggregate), measured in Mt CO_2 emissions

Source: Data from World Bank, 2019

of time. Instead, we see a huge increase in global carbon emissions, equivalent to an increase by about 3.5-fold between 1960 and 2015. This is due to several factors, such as increased population growth, increased economic growth, increased economic output, increased consumption of goods and services, increased energy use, etc.

For the OECD, we can see a decline in absolute emissions for several years following the global financial crisis in 2008. This shows how directly correlated economic growth and emissions growth are. It remains to be seen whether the downturn or stabilising of emissions that has been observed in recent years means a long-term decline or stagnation over longer periods.

Figure 5.6 shows an absolute decoupling of economic growth from emissions as carbon emissions have indeed been reduced in total terms in the European Union (analysed as an aggregate, not as individual countries) between 1960 and 2015. The grey circles and the arrows indicate specific years that have been turning points, which were triggered by historic events such as the oil crisis in the 1970s, the collapse of the Eastern Bloc and the global financial crisis. These three historic events have led to climate change mitigation action that has indeed reduced emissions in the long term. These are the following processes of restructuring economies and industries:

- economic restructuring away from energy- and carbon-intensive industries, partly triggered by the collapse of the Eastern Bloc (e.g. Germany, Eastern Europe) and other large-scale economic restructuring towards the service economy (e.g. UK)
- energy transitions to low carbon energy, e.g. France, Sweden, Denmark
- lower growth/no growth due to the financial crisis, which is fiercely debated and value-laden

Other processes that have led to increased mitigation of emissions in the EU are the outsourcing and offshoring of embedded emissions to Asia and elsewhere and regulatory pressure from the UNFCCC climate policy regime and regional EU decarbonisation policies. Yet, it is difficult to pinpoint these processes to key dates such as the global financial crisis beginning in 2008 or the fall of the Berlin Wall in 1989. These processes and strategies show that it is indeed possible to have economic growth while at the same time reducing emissions. Yet, this requires the large-scale introduction of low carbon energy sources, increases in energy efficiency, economic restructuring of industries and stringent energy and climate policies. Population growth also plays a major role in the increase in energy use and emissions. The major bottlenecks for future decarbonisation are in the restructuring of energy-related activities and industrial change in the transport and industry sectors, as well as with relation to population growth in some rapidly growing countries. These changes need to be accompanied by financial, political and social support for low carbon energy transitions and economy-wide decarbonisation. Finally, while countries such as Sweden and Denmark serve as major front-runners and examples for successful low carbon energy transitions, their emissions are low in comparison to the world's largest emitters, such as China, the USA, India, Russia and Japan (IEA, 2019). It is these major emitters where rapid and large-scale changes are most urgently needed.

Exercises

1a Go to the statistics website of the World Bank: https://data.worldbank.org. Look at the indicator 'access to clean fuels and technologies for cooking' over time (e.g. 1970s to today), both at the worldwide level and for two developing countries of your choice. What dynamics can you see? Has the access to clean cooking fuels increased or decreased over time? Is this development similar or different for the two countries of your choice and at the global level? What does this tell us about the energy ladder?

1b Go to the statistics website of the International Energy Agency IEA: www.iea. org/statistics. Look at the indicator 'energy consumption' for one low-income country, e.g. Tanzania, and one high-income country, e.g. France. Look at the share of electricity that is being used and the share of biofuels and waste (in low-income countries often representing traditional biomass). What are the differences between these two countries? What does this tell us about the energy ladder and fuel switching as countries develop?

2 Use data from the World Bank and the IEA databases to find out the following: (a) the GDP (in current USD) worldwide, the GDP of a chosen high-income country (e.g. Australia or for a contrast Sweden) and the GDP of a chosen low/middle-income country (e.g. Nepal) over time. (b) Look at the CO_2 emissions over time worldwide and for the high- and low/middle-income country of your choice. (c) Look at the energy use over time worldwide and for the high- and low/middle-income country of your choice. What trends do you see with regards to GDP, emissions and energy use? Do you see any decoupling of economic growth from (b) emissions and (c) energy use? If not, why? If yes, how was this achieved? (d) You can analyse the data from the viewpoint of absolute decoupling, such as emission reductions measured in CO_2 emissions in kt, or relative decoupling, such as emission reductions measured in intensity. What differences do you see now? (e) What opportunities and challenges occur for the low-income country concerning decoupling economic growth from emissions/energy use? What can they (not) learn from high-income countries?

References

Barnes, D.F. & Floor, W.M. (1996) Rural energy in developing countries: A challenge for economic development. *Annual Review of Energy and the Environment*, 21 (1), 497–530.

Barrett, M., Lowe, R., Oreszczyn, T. & Steadman, P. (2008) How to support growth with less energy. *Energy Policy*, 36 (12), 4592–4599.

Bhattacharyya, S.C. (2006a) Energy access problem of the poor in India: Is rural electrification a remedy? *Energy Policy*, 34 (18), 3387–3397.

Bhattacharyya, S.C. (2006b) Renewable energy and the poor: Niche or nexus? *Energy Policy*, 34 (6), 659–663.

Boyd, E., Hultman, N., Roberts, Corbera, E., Cole, J., Bozmoski, A., Ebeling, J., Tippman, R., Mann, P., Brown, K. & Liverman, D.M. (2009) Reforming the CDM for sustainable development: Lessons learned and policy futures. *Environmental Science & Policy*, 12 (7), 820–831.

Byrne, R., Ockwell, D., Urama, K., Ozor, N., Kirumba, E., Ely, A., Becker, S. & Gollwitzer, L. (2014) *Sustainable Energy for Whom? Governing Pro-poor, Low Carbon Pathways to Development: Lessons from Solar PV in Kenya.* Brighton, STEPS Centre. STEPS Working Paper 61.

Deininger, K. & Squire, L. (1998) New ways of looking at old issues: Inequality and growth. *Journal of Development Economics*, 57 (2), 259–287.

Drupp, M.A. (2011) Does the Gold Standard label hold its promise in delivering high sustainable development benefits? A multi-criteria comparison of CDM projects. *Energy Policy*, 39 (3), 1213–1227.

Fields, G. (2001) *Distribution and Development, A New Look at the Developing World.* New York, Russell Sage Foundation and Cambridge, Massachusetts, and London, The MIT Press.

Glomm, G. (1997) Whatever happened to the Kuznets curve? Is it really upside down? *Journal of Income Distribution*, 7 (1), 63–87.

Goldemberg, J. & Lucon, O. (2009) *Energy, Environment and Development.* 2nd edition. Oxon, Earthscan, Routledge.

Government of South Africa (2008) *South Africa's Long-term Mitigation Scenarios.* Department of Environment Affairs and Tourism, South Africa. Available from: www.erc.uct. ac.za/sites/default/files/image_tool/images/119/Papers-2007/07Scenario_teamLTMS_ Scenarios.pdf

Heltberg, R. (2004) Fuel switching: Evidence from eight developing countries. *Energy Economics*, 26 (5), 869–887.

Holdren, J.P. & Smith, K.R. (2000) *Energy, the Environment and Health.* World Energy Assessment, Energy and the Challenge of Sustainability. New York, United Nations Development Programme (UNDP).

IEA. (2019) *Statistics.* Paris, International Energy Agency (IEA), OECD/IEA. Available from: www.iea.org/statistics

IEA. (2007) *World Energy Outlook 2007.* Paris, International Energy Agency (IEA), OECD/ IEA.

IPCC. (2007) *Climate Change 2007.* Synthesis Report. Contribution of Working Groups I, II and III to the Fourth Assessment Report of the Intergovernmental Panel on Climate Change (IPCC). [Core Writing Team, Pachauri, R.K and Reisinger, A. (Eds.)]. IPCC, Geneva, Switzerland, 104 pp. Available from: www.ipcc.ch/publications_and_data/ publications_ ipcc_fourth_assessment_report_synthesis_report.htm

Jackson, T. (2009) *Prosperity Without Growth? The Transition to a Sustainable Economy. Sustainable Development Commission.* Available from: http://epubs.surrey.ac.uk/745916

Karp, L. & Liu, X. (2000) *The Clean Development Mechanism and Its Controversies.* Berkeley, University of California. CUDARE Working Paper 903. Available from: http://escholarship. org/uc/item/9739314q

Kuznets, S. (1955) Economic growth and income inequality. *The American Economic Review*, 45 (1), 1–28.

Masera, O.R., Saatkamp, B.D. & Kammen, D.M. (2000) From linear fuel switching to multiple cooking strategies: A critique and alternative to the Energy Ladder Model. *World Development*, 28 (12), 2083–2103.

Mirza, B. & Kemp, R. (2011) Why the rich remain energy poor. Consilience. *The Journal of Sustainable Development,* 6 (1), 133–155. Available from: https://consiliencejournal.org / article/why-rural-rich-remain-energy-poor/

Nussbaumer, P. (2009) On the contribution of labelled Certified Emission Reductions to sustainable development: A multi-criteria evaluation of CDM projects. *Energy Policy*, 37, 91–101.

Ockwell, D.G., Watson, J., MacKerron, G., Pal, P. & Yamin, F. (2008) Key policy considerations for facilitating low carbon technology transfer to developing countries. *Energy Policy*, 36 (11), 4104–4115.

Project Catalyst (2008*) Low-carbon Growth: A Potential Path for Mexico*. Centro Mario Molina.

Saith, A. (1983) Development and distribution: A critique of the cross-country U hypothesis. *Journal of Development Economics*, 13 (3), 367–382.

Srinkanth, S. & Lloyd, B. (2011) Can the Clean Development Mechanism (CDM) deliver? *Energy Policy*, 39 (3), 1600–1611.

Sutter, C. & Parreño, J.C. (2007) Does the current Clean Development Mechanism (CDM) deliver its sustainable development claim? An analysis of officially registered CDM projects. *Climatic Change*, 84, 75–90.

Urban, F. (2010) Pro-poor low carbon development and the role of growth. *International Journal of Green Economics*, 4 (1), 82–93.

Urban, F. & Nordensvärd, J. (2013) *Low Carbon Development: Key Issues*. Oxon, Earthscan, Routledge.

Urban, F. & Nordensvärd, J. (2018) Low carbon energy transitions in the Nordic countries: Evidence from the environmental Kuznets Curve. *Energies*, 11(9), 2209. doi:10.3390/en11092209

Urban, F., Benders, R.M.J. & Moll, H.C. (2009) Energy for rural India. *Applied Energy*, 86, 47–57.

Van Ruijven, B., Urban, F., Benders, R.M.J., Moll, H.C., Van der Sluijs, J., De Vries, B. & Van Vuuren, D.P. (2008) Modeling energy and development: An evaluation of models and concepts. *World Development*, 36 (12), 2801–2821.

World Bank (2008) *Low Carbon Growth in India*. Washington DC, The World Bank.

World Bank (2019) *Data*. Washington DC, The World Bank. Available from: https://data.worldbank.org/

6

THE ENERGY–POVERTY–CLIMATE NEXUS

Linkages between energy, poverty and climate

High levels of poverty are common in countries across sub-Saharan Africa, developing Asia and parts of Latin America, which partly coincides with the hotspots of energy poverty. In sub-Saharan Africa, almost half of the population does not have access to electricity and about 80% rely on traditional biomass (IEA, 2019). Electrification rates are even lower in rural areas.

We discussed earlier that about 1 billion people worldwide do not have access to electricity and 2.7 billion people rely on traditional biomass – such as fuelwood and dung – for basic needs such as cooking and heating (IEA, 2019). These people are considered as living in energy poverty. Energy poverty is one of the many facets of poverty. Poverty alleviation depends on increased access to services, goods and information (Casillas & Kammen, 2010). We saw in Chapter 4 that providing energy access is a prerequisite for achieving the UN's Sustainable Development Goals (SDGs). Poverty alleviation and achieving the SDGs are challenged by lack of access to reliable, modern energy services and climate change. Casillas and Kammen (2010: 1181) suggest that 'mitigating climate change, increasing energy access and alleviating rural poverty can be complementary'. The same can be said for urban and peri-urban poverty too. The link between energy, climate change and poverty is called the energy–poverty–climate nexus.

Energy access alone is not a panacea for eradicating poverty. However, it can have immediate effects for the socioeconomic well-being and development of those living in poverty and have widespread positive environmental effects. It can thereby contribute to poverty eradication and, as we said above, energy access is a prerequisite for achieving other development goals.

The energy and poverty link

Due to energy poverty, basic needs are not met and socioeconomic well-being remains low. Adverse health impacts due to energy poverty are high, economic activity tends to remain low and educational opportunities may not be met. This is particularly the case for women and girls. Access to electricity and modern fuels stimulates economic activity, can have health benefits and can increase opportunities for education and meeting the demands of businesses and households through increasing the availability, quality and often also the quantity of energy services (Casillas & Kammen, 2010).

The climate change and poverty link

While climate change has mainly been caused by today's industrialised countries, it affects the world's poorest the worst. This is due to their high vulnerability and their limited resources to adapt to climatic impacts. Climate change can exacerbate existing risks, such as water stress, droughts and pressure on food security in drought-prone areas, or flooding, storms and risks from sea-level rise in low-lying coastal areas (Casillas & Kammen, 2010). The majority of people currently affected ('at risk') from climatic impacts and those at risk from sea-level rise happen to live in areas that are also considered global poverty hotspots.

The energy–poverty–climate nexus

Access to reliable, affordable, modern energy can contribute to alleviating energy poverty, thereby contributing to alleviating poverty in general as well as contributing to mitigating and adapting to climate change. With regard to climate change adaptation, modern energy can help build the resilience of individuals, households and communities by contributing to improving socioeconomic capacities to adapt to a changing climate and increase their resilience to climatic shocks; for example, by improving health and generating additional income. With regard to mitigating GHG emissions that lead to climate change, the world's poor have very low emissions. However, GHG emissions from countries that currently have low energy access rates are likely to grow in the future. This is partly based on the prevalence of and preference for decentralised diesel generators as an option for creating access to modern energy. However, this tends to result in dependency on expensive, often imported, high carbon fuels (Casillas & Kammen, 2010). Casillas and Kammen (2010) used a marginal abatement cost curve to calculate the annual carbon abatement potential for technological interventions, mostly in the field of energy efficiency and low carbon energy, and the costs of mitigation. They found that for rural areas in Nicaragua, the cheapest and most carbon-saving technologies were basic energy efficiency measures such as installing meters and energy-efficient lighting, as well as introducing biogas for cooking and wind turbines for electricity generation.

They calculated that 'every dollar spent on the transition to more efficient low carbon energy systems in rural areas has the potential to produce greater human development, savings, and carbon mitigation returns than in more industrialized areas' (Casillas & Kammen, 2010: 1181). Even in urban and peri-urban areas, low carbon energy can contribute to human development, cost savings and carbon mitigation (Urban, 2014).

The political economy of energy poverty

This section discusses the political economy of energy poverty and how it links to development and climate change. The SE4All initiative aims to provide modern energy access to everyone worldwide by 2030, reduce global energy intensity by 40% and increase the share of renewable energy among global primary energy supply to 30% (Sovacool, 2013). Despite being aware of the global social, environmental and economic implications of energy poverty, it is still widespread, and eradicating energy poverty has not received sufficient attention in the last decades. Sovacool (2012) explains this in political economy terms. The political economy could be roughly defined as a social science approach that is concerned with the politics of economics. This relates to how a country's economy is being governed, taking into account political and economic factors. At the core of political economy are the 'relationships between political power and interests on one hand and economic power and interests on the other' (Lockwood, 2013: 26). There are several political-economy-related reasons why the alleviation of energy poverty is slow and energy access stagnates:

- Private energy utilities are often reluctant to invest in expanding grids and energy infrastructure to areas that have no access to modern energy. This is due to high and often risky investments where returns are not guaranteed, as those living in energy poverty often have low incomes, a low ability to pay and low levels of energy consumption. This often goes along with electricity theft and high operation and maintenance losses (Sovacool, 2012; Urban, 2014).
- Public sector actors often have limited budgets and competing interests (e.g. money available for schools/hospitals *or* grids, not both). They also often depend on public opinion and election outcomes, often tend to work to election cycles and may tend to consider energy projects as a charitable act rather than attracting votes. Governments may also not understand the energy needs of the poor (Sovacool, 2012).
- Historically, support for urban areas has been stronger than for rural areas with regard to energy access. Rural energy demand is often dispersed, requiring decentralised or off-grid technology, such as renewable energy technology, rather than centralised grid extensions (Sovacool, 2012).
- In many poor countries, there is a lack of access to modern energy technology, particularly decentralised or off-grid technology, such as renewable energy

technology. Access often depends on technology transfer from wealthy donor countries. This can result in countries aiming to get overseas market access or trade agreements or conduct pilot tests, rather than supporting domestic modern energy access needs (Sovacool, 2012).

- There is a risk that providing modern energy for basic needs, such as electric lighting or improved cooking stoves, does not address more fundamental poverty and well-being issues (Sovacool, 2012).
- Other issues have an effect on the uptake of modern energy technology (Sovacool, 2012). These include the limited ability of households to pay for energy services and equipment, electrification at the village level rather than at the household level and expensive connection fees for accessing the grid for individual households, and cultural issues, such as treating solar hot water heaters as risky or biogas as impure.
- Other factors include lengthy planning approvals, non-transparent and complicated processes for renewable energy projects and lack of maintenance for energy projects (Sovacool, 2012). Another reason why inequalities in energy access still prevail is linked to fossil fuel subsidies. Lockwood (2013) refers to fossil fuel subsidies as 'negative carbon prices'. He suggests that:

> Removing or reforming such subsidies would in many cases both reduce carbon emissions and free up resources for spending on services for poor people (World Bank, 2010). For example, in some Indian states, as much as 50% of the budget goes on subsidies to electricity (Joseph, 2010). Moreover, most of the benefits from fuel subsidies are captured by the middle class rather than the poor. For example, the International Energy Agency estimates that of US\$22.5 billion spent by India on fossil fuel subsidies in 2010, less than US\$2 billion benefited the poorest 20 per cent of the population (IEA, 2011: p. 40). However, attempts to remove subsidies typically meet strong opposition, often from poorer urban populations. This is partly because poor people are proportionately more affected by subsidy removal. But it also because there is little confidence that money saved by reduced subsidies will be spent on pro-poor provision of public goods in what are often highly corrupt countries. In the face of such opposition it is politically risky for regimes to cut subsidies.
>
> *Lockwood (2013: 27)*

Challenges for development

This section discusses the challenges for development and poverty reduction that arise from the energy–poverty–climate nexus. A rise in fossil fuel energy access can be linked to adverse environmental impacts, such as increased GHG emissions that contribute to climate change. This dynamic will be debated from the perspective of continued energy poverty and carbon lock-in.

Challenge: continued energy poverty

Energy poverty has existed for the whole of human history. It is a fundamental struggle for humans. In recent decades, large sums of money have been invested in rural and urban electrification programmes, and large improvements have been made in Organisation for Economic Co-operation and Development (OECD) countries as well as elsewhere, for example, in China. In recent years, the phenomenon of energy poverty has received more attention worldwide, partly thanks to the UN's SE4All initiative. Nevertheless, energy poverty continues to prevail in many areas, particularly in sub-Saharan Africa and developing Asia. This is partly because some of the efforts of governments, donor agencies, non-governmental organisations (NGOs) and the private sector are being 'outnumbered' by population growth, inequitable distribution of wealth and resources, as well as structural problems such as underfunded utilities, outdated infrastructure, the dispersed nature of those living in energy poverty and immense sums of funding required. Other key areas may be prioritised by governments, donor agencies and NGOs, such as food security, peace building, health and education. Again, other factors related to wars, violent conflict, political instability and corruption undermine adequate investment in new energy infrastructure and projects. In addition, natural disasters and extreme weather events can damage or destroy existing energy infrastructure. These factors contribute to poverty, of which energy poverty is just one specific element.

Due to the complexity and scale of these human development dilemmas, there is a very real risk that energy poverty may continue to exist for several (if not many) decades to come. The International Energy Agency (IEA) estimates that without major intervention, there would still be more than 1 billion people without access to electricity in 2030 (similar to today) and 2.6 billion without access to clean cooking fuels in 2030 (compared with 2.7 billion today), of whom about 85% would live in rural areas (IEA, 2019). The IEA aims for universal energy access for everyone worldwide by 2030; nevertheless, the reality of achieving these targets is still far off. The IEA estimates that nearly US$1 trillion in cumulative investment is needed to achieve universal energy access by 2030 (IEA, 2011).

Challenge: development with carbon lock-in

We discussed in the section 'The energy–poverty–climate nexus' that energy access is a prerequisite to alleviating poverty and achieving the SDGs. Hence, poverty alleviation will only be possible if access to energy is increased for the world's poor, particularly for the rural poor who are often cut off from energy infrastructure and energy markets. At the same time, a rise in fossil fuel energy access can be linked to adverse environmental impacts, such as increased GHG emissions. An analysis from India on universal electrification by 2030 for rural households that do not have access to electricity evaluated three different modes of providing electricity access: grid-based, decentralised with renewable energy technology and decentralised with diesel generators. The renewable energy technology assessed includes solar photo-voltaic (PV), micro-wind, micro-hydro, biogas-based electricity and stand-alone renewables such as biogas cookers, solar cookers, solar water heaters and solar lamps (Urban, 2014). The study by Urban et al. (2009) is in line with the IEA's study on how to achieve the UN universal energy access target by 2030. The IEA aims that, by 2030, 100% of all households should have access to electricity. In rural areas, this is likely to involve 30% grid connections and 70% decentralised mini-grid and stand-alone off-grid renewable energy options (IEA, 2010). The rural India study finds that the decentralised electricity access option with renewable energy technology could save nearly 100% of all GHG emissions compared with grid connections and diesel generators. It also reduces overall primary energy use compared with the grid and diesel generators and is cost effective in comparison to diesel generators and grid connections that are farther away (Urban et al., 2009; Urban, 2014). For rural India, the transmission costs of centralised grid power and decentralised PV-power are about equal at a distance of around 10 km to the nearest grid, while the transmission costs double for grid power at a distance of about 25 km (Chakrabarti & Chakrabarti, 2002; Dijkstra, 2007). Other research also indicates that decentralised renewable energy can be financially attractive compared with conventional centralised grid power (Sinha & Kandpal, 1991; Abe et al., 2007; Nouni et al., 2008). This is due to low load factors, long distribution lines and high associated transmission and distribution losses of centralised grid power (Sinha & Kandpal, 1991). Achieving universal electrification by 2030 for rural Indian households comes with a main challenge: carbon lock-in. Carbon lock-in plays a role as the investments and the infrastructure for energy may be tied to high carbon pathways for decades due to the long construction times and lifetimes of fossil fuel power plants, as well as the high investments and running costs for fossil fuel power plants. In the case of India, the IEA (2007) estimates that between 2006 and 2030, US$956 billion will need to be invested in new power infrastructure alone. However, investment costs usually tend to make up less than one third of the total costs of running fossil power plants. The remaining two thirds of the costs are due to operation and maintenance costs (O&M) and fuel costs. The study by Urban et al. (2009) and Urban (2014) on universal electrification by 2030 for Indian rural households compared the whole systems costs (investment costs to be paid by investors, the O&M costs to be paid by utilities and the fuel costs to be paid both by utilities and by private households) for the three different modes of providing electricity access: grid-based, decentralised with renewable energy technology and decentralised with diesel generators. They calculated that for the period 2005 to 2030, the whole system costs for diesel-powered electrification would be about 2.5 times more expensive than grid connections and between 1.4 times to almost 10 times more expensive than renewable-energy-powered electrification, depending on choice of off-grid versus mini-grids and types of renewable energy technology. With regard to carbon lock-in and carbon emissions, rural Indian households have very low energy consumption, and this is only expected to change slowly once access to electricity is provided. This is due to limited incomes, expenditure and energy infrastructure. The rural electrification study found that once electricity access is provided, rural households might add between another 25% (grid-based) up to 40% (diesel-based) to India's total carbon dioxide emissions per year by 2030. This is a considerable amount that would have an overall effect on India's carbon emissions and global emissions. The emissions would be even higher if higher consumption levels were attained in rural households. Providing universal electrification in rural India with decentralised renewable energy was found to have the lowest carbon dioxide emissions and thus the highest climate change mitigation potential (Urban et al., 2009; Urban, 2014).

These findings highlight the challenges to the energy–poverty–climate nexus. It should be the unquestionable right of every person in the world to have access to clean, affordable, modern energy. Access to electricity and modern energy for cooking should therefore be treated as a universal right. Nevertheless, this opens up the challenges described above, namely that this may lead to a significant rise in national and global carbon emissions. However, this does not need to be the case. Instead, the energy–poverty–climate nexus offers opportunities for sustainable development in a carbon-constrained world, by favouring renewable energy options.

Opportunities for development

This section discusses the opportunities for development and poverty reduction that arise from the energy–poverty–climate nexus. A rise in renewable energy access can contribute to climate change mitigation. Mitigating climate change, increasing energy access and alleviating poverty can therefore be complementary (Casillas & Kammen, 2010). This section explores two options for development, namely low carbon development and low carbon climate-resilient development.

Opportunity: low carbon development

Low carbon development refers to:

> a development model that is based on climate-friendly low carbon energy and follows principles of sustainable development, makes a contribution to avoiding dangerous climate change and adopts patterns of low carbon consumption and production (Skea & Nishioka, 2008; Urban, 2010; Urban et al., 2011a).
>
> *Urban and Nordensvärd (2013: 5)*

Some may phrase low carbon development more along the terms of economic growth:

> In growth terms, low carbon development is defined as using less carbon for growth, which includes:
>
> – switching from fossil fuels to low carbon energy
> – promoting low carbon technology innovation and business models

- protecting and promoting natural carbon sinks such as forests and wetlands
- formulating policies that promote low carbon practices and behaviours (DFID, 2009: 58; Urban et al., 2011a).

The ultimate aim of low carbon development is to mitigate emissions to avoid dangerous climate change while at the same time achieving social and economic development (Urban et al., 2011b).

Urban and Nordensvärd (2013: 5–6)

Some rather refer to 'low emissions development' than to low carbon development. This takes into account GHGs, such as methane and nitrous oxide from agriculture and land-use change, other than carbon dioxide. This book uses the term low carbon development, as carbon dioxide is the main GHG and we deal with GHG emissions from energy. Figure 6.1 shows how climate change mitigation and human development goals overlap to form the concept of low carbon development.

Low carbon development can bring opportunities and benefits for high-income, middle-income and low-income countries.

In low-income and lower middle-income countries, issues of social justice and poverty reduction are the key to low carbon development, while for higher middle-income and high-income countries low carbon innovation and emission reductions are at the heart of implementing low carbon development.

(Urban, 2013: 11)

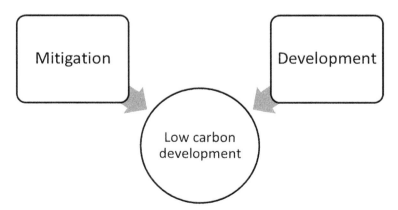

FIGURE 6.1   Low carbon development (LCD): the interface between climate change mitigation and development

Source: Amended from Urban and Nordensvärd (2013: 7)

The benefits of low carbon development for poor countries and poor people are linking into the energy–poverty–climate nexus. These benefits include, for example:

- providing access to climate-friendly modern energy, such as electricity from PV or biogas, for lighting, cooking, heating and other basic needs, as an alternative to traditional fuels and fossil fuels
- providing low carbon energy for income-generating activities and educational purposes (e.g. solar-powered mobile phone charging business or wind-powered electricity for lighting a school)
- increasing energy security
- increasing energy access
- social benefits, such as improved health, for example, through reduced indoor air pollution when switching from fuelwood to modern energy options
- opportunities for green job creation, such as jobs in the renewable energy sector
- accessing low carbon technology and innovation
- avoiding carbon lock-in, which means avoiding that investments and infrastructure are locked into high carbon pathways for decades
- creating political support for the climate change negotiations and other diplomatic issues

(Urban, 2013: 1–12)

Low carbon development offers opportunities for developing countries to avoid the 'dirty path' of development which today's industrialised countries follow and instead develop more sustainably by using the latest climate- and environment-friendly technologies (technological leapfrogging).

Innovation and leapfrogging might also lead to 'jumping over' potentially polluting periods, like the coal era during the Industrial Revolution. Developing countries may be able to avoid such periods due to cleaner and more advanced low carbon technologies (Urban, 2014). Low carbon development can therefore offer two opportunities: to overcome energy poverty by switching from traditional biomass to modern fuels and to reduce emissions by choosing renewable energy rather than fossil fuels. For example, households and individuals living in energy poverty can switch from fuelwood to cooking on biogas or solar cookers and thereby reduce negative health impacts from indoor air pollution. For lighting, households and individuals can switch from kerosene lamps to solar lamps and thereby reduce the dangers of kerosene lamps, increase the time available for studying after dark, reduce dependency on expensive kerosene and reduce their emissions. Those living in energy poverty may even be able to get electricity access from solar PV, micro-hydro plants or micro-wind turbines, which can enable them to study after dark, have electricity for income-generating purposes and a wide range of other purposes, such as refrigeration, fans, radios and mobile phone charging, as well as reduce adverse health impacts, the burden of fuelwood collection and reduce emissions. This could also help to avoid a future carbon lock-in and increase energy security as indigenous resources that are abundantly available, such as the sun and wind, are being used rather than costly, imported fossil fuels. Low carbon development can therefore contribute to alleviating energy poverty and contribute to sustainable development in a changing climate. Yet, access to these technologies depends on access to funding, markets, infrastructure and services which can all be bottlenecks.

There are also more radical ideas about low carbon development and climate change more generally, which are embedded in wider issues on natural resource management, such as distributing resources more equally, owning natural resources collectively and managing them cooperatively, which could result 'in an improved welfare, a better quality of life and greater democratic control of production and (renewable) resources' (Storm, 2009: 1026; Nordensvärd, 2013: 72). Jackson (2009) criticises the notion of continuing economic growth around the world and argues that there is no socially just, environmentally sustainable world that can sustain 9 billion people with continually growing incomes. He therefore refers to 'prosperity without growth' or prosperity within the environmental limits of our planet. De-growth would need to focus on social well-being, quality of life and environmental sustainability. Nordensvärd (2013) mentions that:

> the state would need to ensure that economic resources are more evenly distributed so that the poor would not be disproportionally affected. This might imply sharing some of the wealth of nations more equally, such as by taxing the richer groups of society more to ensure that the poorer groups of society are well off despite a de-growth or reducing the excessive pay of top firm executives and creating employment opportunities with the available funding. . . . Understandably, in our current economic climate these are controversial ideas that are rarely well-received by the powerful and wealthy. . . . Some scholars go even further and argue that capitalism's search for growth and profit is not compatible with ecological sustainability. Max Weber discusses in his famous book *The Protestant Ethic: The Spirit of Capitalism* that the capitalist economic world order will proceed until 'the last ton of fossilized coal is burnt' (Weber, 1953: 181). Scholars such as Bello (2008) . . . argue that we are heading towards either a collapse of the present capitalist system or a collapse of our global climate.
>
> *Nordensvärd (2013: 72)*

Schweickart (2008) suggests that we must move beyond capitalism to enable sustainable development. While these are very outspoken views, others such as Newell (2011: 4) discuss the 'sound of silence' that most scholars have towards capitalism and its relationship to global environmental change. Whichever political view one has, it is clear that efforts towards low carbon development could help tackle the challenges posed by climate change, energy poverty and natural resource depletion. Nevertheless, low carbon development is not a technical fix. It involves a complex set of social, cultural, political, economic, technological and geographic conditions to achieve low carbon development. Urban and Nordensvärd (2013) therefore write:

> Low carbon development can bring opportunities and benefits for both developed and developing countries, nevertheless low carbon development can only be implemented when an adequate enabling environment is in place which addresses the political, economic, social and technological key issues. However, the greatest challenge of all is to overcome the current mindset and to develop alternative, more sustainable and more equitable development models for humankind.
>
> *Urban and Nordensvärd (2013: 321)*

Opportunity: low carbon, climate-resilient development (aka climate-compatible development)

In the previous section, we discussed what low carbon development is and how it can contribute to alleviating energy poverty and to sustainable development in a changing climate. While this addresses the concerns of climate change mitigation and development, climate change adaptation has been largely left out of the equation. However, it is of utmost importance that individuals, households, communities and countries adapt to a changing climate and build up their resilience. This is the notion of low carbon climate-resilient development which we will discuss here. Please note that the terms climate-compatible development and low carbon climate-resilient development are used interchangeably in the literature, with the first term being more often used by practitioners and the second term being more often used by academics. Low carbon climate-resilient development aims to deliver lower emissions development that is resilient to current and future climatic impacts and can be defined as 'development that minimises the harm caused by climate impacts, while maximising the many human development opportunities presented by low emissions, more resilient, future' (Mitchell & Maxwell, 2010: 1). It is a concept that emerged about a decade ago as the study of the synergies or co-benefits of climate change adaptation and mitigation became more popular (Klein et al., 2005, 2007) and as development concerns were integrated into this agenda. Low carbon climate-resilient development is therefore at the interface of mitigation, adaptation and human development. Figure 6.2 shows how climate change mitigation, adaptation and human development goals overlap to form the concept of low carbon climate-resilient development/climate-compatible development.

While climate change adaptation and mitigation are often treated as separate issues by development agencies and academia, they are in fact closely intertwined in the lives of people that are faced with the impacts of climate change. Treating climate change as one holistic challenge, rather than two separate ones, can therefore help to tackle the causes and impacts of climate change in an integrated way and address the synergies (Urban, 2013). This also has the potential to deal with the energy–poverty–climate nexus by contributing to alleviating energy poverty,

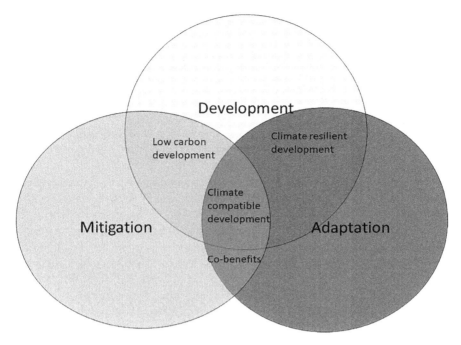

FIGURE 6.2 Low carbon climate-resilient development/climate-compatible development: the interface between climate change mitigation, adaptation and development

Source: Redrawn from Urban and Nordensvärd (2013: 7), adapted from Mitchell and Maxwell (2010)

reducing emissions and aiming for sustainable development. We will now discuss one example of low carbon climate-resilient development.

The solar-powered pump project for irrigation in Benin

Benin is a country in sub-Saharan Africa, more specifically in West Africa. It is a low-income country, with a yearly per capita income (measured as GDP per capita PPP) of only US$2,510 in 2018 (compared with US$55,680 in the United States) (World Bank, 2019). The country depends heavily on agriculture; nevertheless, climatic impacts have made it more prone to water stress and drought, which makes agriculture difficult, especially in the dry season. At the same time, the country receives abundant sunshine, which makes it ideal for using solar panels for electricity for irrigation. This is why the solar-powered pump project for irrigation in the Kalalé district of rural Benin is useful in promoting solar-powered irrigation for agriculture. For the project, Stanford University and the Solar Electric Light Fund worked together to set up an irrigation system for agriculture based on solar PV systems in Benin (SELF, 2008). The PV systems power a pump and a drip-irrigation system, which provides regular access to water for the agricultural activities of the villagers. It enables the villagers to grow agricultural produce, mainly vegetables, for their own use and to sell to local markets, even in the dry season. At the same time, the PV systems provide electricity for local community buildings such as schools and clinics. The renewable energy systems replace diesel generators for water pumping and electricity generation, thereby avoiding carbon emissions. At the same time, the PV systems improve access to water and irrigation for farmers and households in a changing, more drought-prone climate. In addition, they enable farmers and the community to make a modest income from selling their agricultural produce on the market and thereby improve their livelihoods (SELF, 2008). Hence, there are mitigation, adaptation and development aspects (Urban, 2013: 208).

Other examples of low carbon climate-resilient development are mangrove afforestation programmes in coastal Vietnam that contribute to carbon storage, coastal protection and livelihoods from seafood harvesting and selling (Urban, 2013).

However, not every project or intervention that carries the low carbon climate-resilience label contributes adequately to mitigation, adaptation and development. There is a risk that projects aim to address all three complex issues, with the same capacities and financial means available for one of these issues, which might result in poorer outcomes. In addition, governments may implement low carbon climate-resilient development projects because of donor pressure rather than because of national needs and priorities (Hagemann et al., 2012). There may also be trade-offs, such as with the development of biofuels, which can have negative impacts on food security and land rights of local communities and indigenous people (Urban, 2013). Despite these challenges, there is the potential that low carbon climate-resilient development can contribute to solving the energy–poverty–climate nexus by alleviating energy poverty while mitigating GHG emissions and increasing adaptive capacity to climatic impacts.

Exercises

1 Imagine you are working for a government authority, an NGO or a business in a developing country. You are giving a speech at a workshop that is being organised by a major development agency that would like to fund activities in your country. How would you explain the energy–poverty–climate nexus to them? What kind of funding for which projects (that relate to the energy–poverty–climate nexus) would be useful for your country? How could you convince the funders that they should invest in these? You may wish to prepare a short speech or a short PowerPoint presentation.

2 Can you think of any practical examples of low carbon, climate-resilient development interventions? You might want to do some research online.

References

Abe, H., Katayama, A., Bhuwneshwar, P.S., Toriu, T., Samy, S., Pheach, P., Adams, M.A. & Grierson, P.F. (2007) Potential for rural electrification based on biomass gasification in Cambodia. *Biomass and Bioenergy*, 31 (9), 656–664.

Bello, W. (8 April 2008) *Will Capitalism Survive Climate Change?* ZNet. Available from: http://zcomm.org/znetarticle/will-capitalism-survive-climate-change-by-walden-bello/

Boyd, E., Hultman, N., Roberts, Corbera, E., Cole, J., Bozmoski, A., Ebeling, J., Tippman, R., Mann, P., Brown, K. & Liverman, D.M. (2009) Reforming the CDM for sustainable development: Lessons learned and policy futures. *Environmental Science and Policy*, 12 (7), 820–831.

Casillas, C. & Kammen, D. (2010) The energy–poverty–climate nexus. *Science*, 26 (330), 1181–1182.

Chakrabarti, S. & Chakrabarti, S. (2002) Rural electrification programme with solar energy in remote region – a case study in an island. *Energy Policy*, 30 (1), 33–42.

DFID. (2009) *Eliminating World Poverty: Building our Common Future*. London, Department for International Development DFID. DFID White Paper.

Dijkstra, T. (2007) *Photovoltaic Panels as an Alternative to Fossil Fuel Energy Sources in Rural Areas of India*. Groningen, IVEM, University of Groningen.

Drupp, M.A. (2011) Does the Gold Standard label hold its promise in delivering high Sustainable Development benefits? A multi-criteria comparison of CDM projects. *Energy Policy*, 39 (3), 1213–1227.

Hagemann, M., Harvey, B., Urban, F., Naess, L.O., Höhne, N. & Hendel-Blackford, S. (2012) *Planning Climate Compatible Development: The Role of Tools and Methodologies*. London. Climate and Development Knowledge Network (CDKN).

IEA. (2019) *Statistics*. Paris, International Energy Agency (IEA), OECD/IEA. Available from: www.iea.org/statistics/

IEA. (2007) *World Energy Outlook 2007*. Paris, International Energy Agency (IEA), OECD/IEA.

IEA. (2010) *World Energy Outlook 2010*. Energy Poverty: How to Make Modern Energy Access Universal? Paris, International Energy Agency (IEA), OECD/IEA. Available from: http:// www.worldenergyoutlook.org/media/weowebsite/2010/weo2010_poverty.pdf

IEA. (2011) *World Energy Outlook 2011*. Energy for All. Financing Access for the Poor. Paris, International Energy Agency (IEA), OECD/IEA. Available from: www.worldenergy outlook.org/media/weowebsite/2011/weo2011_energy_for_all.pdf

Jackson, T. (2009) *Prosperity Without Growth*. Oxon, Earthscan.

Jakob, M. & Steckel, J.C. (2014) How climate change mitigation could harm development in poor countries. *WIREs Climate Change*, 5, 161–168.

Joseph, K. (2010) The politics of power: Electricity reform in India. *Energy Policy*, 38, 503–511.

Karp, L. & Liu, X. (2000) *The Clean Development Mechanism and Its Controversies*. University of California, Berkeley. CUDARE Working Paper 903. Available from: http://escholarship.org/uc/item/9739314q

Klein, R.J.T., Huq, S., Denton, F., Downing, T.E., Richels, R.G., Robinson, J.B., Toth, F.L. (2007) Interrelationships between adaptation and mitigation. In: Parry, M.L., Canziani, O.F., Palutikof, J.P., van der Linden, P.J. & Hanson, C.E. (Eds.) *Climate Change 2007: Impacts, Adaptation and Vulnerability. Contribution of Working Group II to the Fourth Assessment Report of the Intergovernmental Panel on Climate Change*. Cambridge University Press, Cambridge, UK. pp. 745–777.

Klein, R.J.T., Schipper, E.L.F. & Dessai, S. (2005) Integrating mitigation and adaptation into climate and development policy: Three research questions. *Environmental Science and Policy*, 8 (6), 579–588.

Lockwood, M. (2013) The political economy of low carbon development. In: Urban, F. & Nordensvärd, J. (Eds.) *Low Carbon Development*. Oxon, Routledge. pp. 25–37.

Mitchell, T. & Maxwell, S. (2010) *Defining Climate Compatible Development*. Climate and Development Knowledge Network (CDKN). Policy Brief November 2010. Available from: www.cdkn.org/wp-content/uploads/2011/02/CDKN-CCD-DIGI-MASTER-19NOV.pdf

Newell, P. (2011) The elephant in the room: Capitalism and global environmental change. *Global Environmental Change*, 21 (1), 4–6.

Nordensvärd, J. (2013) Social policy and low carbon development. In: Urban, F. & Nordensvärd, J. (Eds.) *Low Carbon Development: Key Issues*. Oxon, Routledge.

Nouni, M.R., Mullick, S.C. & Kandpal, T.C. (2008) Providing electricity access to remote areas in India: An approach towards identifying potential areas for decentralised electricity supply. *Renewable and Sustainable Energy Reviews*, 12 (5), 1187–1220.

Nussbaumer, P. (2009) On the contribution of labelled Certified Emission Reductions to sustainable development: A multi-criteria evaluation of CDM projects. *Energy Policy*, 37, 91–101.

Ockwell, D.G., Watson, J., MacKerron, G., Pal, P. & Yamin, F. (2008) Key policy considerations for facilitating low carbon technology transfer to developing countries', *Energy Policy*, 36, 4104–4115.

Schweickart, D. (2008) Is sustainable capitalism possible? *Procedia Social and Behavioral Sciences*, 41 (2010), 6739–6752.

SELF. (2008) Benin Solar Irrigation Project Profile. Solar Electric Light Fund (SELF), Washington DC. Available from: www.stanford.edu/group/solarbenin/references/Benin-ProjectProfile.pdf

Sinha, C.S. & Kandpal, T.C. (1991) Decentralized versus grid electricity for rural India: The economic factors. *Energy Policy*, 19 (5), 441–448.

Skea, J. & Nishioka, S. (2008) Policies and practices for a low carbon society. *Climate Policy, Supplement Modelling Long-Term Scenarios for Low carbon Societies*, 8, 5–16.

Sovacool, B. (2012) The political economy of energy poverty: A review of key challenges. *Energy for Sustainable Development*, 16 (3), 272–282.

Sovacool, B. (2013) A qualitative factor analysis of renewable energy and Sustainable Energy for All (SE4ALL) in the Asia-Pacific. *Energy Policy*, 59 (2), 393–403.

Srinkanth, S. & Lloyd, B. (2011) Can the Clean Development Mechanism (CDM) deliver? *Energy Policy*, 39 (3), 1600–1611.

Storm, S. (2009) Capitalism and climate change: Can the invisible hand adjust the natural thermostat? *Development and Change*, 40 (6), 1011–1038.

Sutter, C. & Parreño, J.C. (2007) Does the current Clean Development Mechanism (CDM) deliver its sustainable development claim? An analysis of officially registered CDM projects. *Climatic Change*, 84, 75–90.

UNFCCC. (1998) *Kyoto Protocol to the United Nations Framework Convention on Climate Change*. United Nations, New York. Available from: http://unfccc.int/resource/docs/convkp/ kpeng.pdf

Urban, F. (2010) Pro-poor low carbon development and the role of growth. *International Journal of Green Economics*, 4 (1), 82–93.

Urban, F. (2013) Triple win? The case of climate compatible development. In: Urban, F. & Nordensvärd, J. (Eds.) *Low Carbon Development: Key Issues*. Oxon, Earthscan, Routledge. pp. 202–212.

Urban, F. (2014) *Low Carbon Transitions for Developing Countries*. Oxon, Earthscan, Routledge.

Urban, F., Benders, R.M.J., Moll, H.C. (2009) Energy for rural India. *Applied Energy*, 86 (Supplement 1), S47–S57.

Urban, F., Mitchell, T. & Silva Villanueva, P. (2011a) Issues at the interface of disaster risk management and low carbon development. *Climate and Development*, 3 (3), 259–279.

Urban, F. & Nordensvärd, J. (2013) Low carbon development: Origins, concepts and key issues. In: Urban, F. & Nordensvärd, J. (Eds.) *Low Carbon Development: Key Issues*. Oxon, Earthscan, Routledge. pp. 3–22.

Urban, F., Watt, R., Ting, M.B., Crawford, G., Wang, Y., Lakew, H., Atta-Owusu, F., Edze, P., Cobson-Cobbold, J.C., Wang, S. & Smith, C. (2011b). *Achieving Low Carbon Development in Low and Middle Income Countries – The Role of Governments, Business and Civil Society*. IDS Project Report for the DFID Learning Hub on Adaptation and Low Carbon Development. Brighton, IDS.

Weber, M. (1953) *The Protestant Ethic and the Spirit of Capitalism*. New York, Scribner.

World Bank (2019) *Open Data*. Washington DC, The World Bank. Available from: http://data.worldbank.org/

World Bank (2010) *Subsidies in the Energy Sector: An Overview*. The World Bank, Washington DC. Background Paper for the World Bank Group Energy Sector Strategy. Available from: http://siteresources.worldbank.org/EXTESC/Resources/Subsidy_background_paper.pdf

7

ENERGY AND CLIMATE POLICY OF MAJOR EMITTERS

Who is responsible for climate change?

Historical responsibility for climate change

As we discussed the energy–poverty–climate nexus in the previous chapter, let us step back for a moment to look into the responsibility for climate change. The development of humankind dates back thousands of years. Even though humans have always exploited nature in some way or another, this development has only become excessively unsustainable during the last two centuries. Many of today's global climate change problems and other environmental problems are assumed to be due to the increased consumption and production of industrialised countries fuelled by fossil fuels since the Industrial Revolution in the late 18th century. This process is continuing today, as developed countries are characterised by industriali-sation, high levels of consumption and production, unsustainable development patterns and a heavy reliance on fossil fuels. It is estimated that about 75% of emissions leading to climate change have historically been emitted by developed countries according to a report by the World Resources Institute published in 2005 (WRI, 2005). Yet, the situation has changed within the last decade as emerging economies like China and India have grown rapidly in terms of the size of their economy, population and also greenhouse gas emissions and energy use (World Bank, 2019). A more recent report by Climate Analytics (2015) based on modelling studies suggests that by 2100, the historical responsibility for global temperature increases will be about 20% for the USA, 17% for the EU, 12% for China, followed by Russia and India.

The largest growth in future energy use and greenhouse gas (GHG) emissions will likely come from emerging economies. It is also estimated that the 50 least developed countries have historically caused only 1% of the emissions leading to climate change (WRI, 2005). Poor people and poor countries are, nevertheless, the most affected by climate change as they lack the financial, technical, infrastructural and institutional resources to adapt.

Current total emissions, per capita and future contribution to climate change

While from a historical perspective climate change is clearly the responsibility of today's developed countries, emerging economies such as China and India are increasingly becoming emerging emitters. Emerging economies have developed within a few decades from relatively modest users of energy to some of the world's major consumers of energy and natural resources. They depend heavily on fossil fuels, particularly coal. Recently, they have become significant emitters of GHG measured in absolute terms, i.e. measured in total emissions. Since 2007, China has overtaken the USA as the world's largest emitter of GHGs in absolute terms. India is currently ranked third in the global top five in terms of the world's largest emitters of GHGs in absolute terms (IEA, 2019). This can be seen in Figure 7.1. However, as an aggregate, the 28 countries of the European Union (EU) come third, after China and the USA (see Figure 7.2). These global emission trajectories are very closely correlated to global energy use trajectories, as China, the USA, the EU, India, Russia and Japan are also the major energy consumers. This can be seen in Figures 7.3 and 7.4.

It is also estimated that emerging emitters such as Brazil, China, India, Indonesia, Turkey and South Africa will account for a large share of future emissions, particularly China, due to large total emissions driven by fast economic growth and large populations. This is one of the reasons why the Paris Agreement included

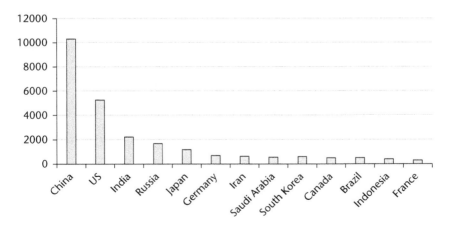

FIGURE 7.1 The world's top CO_2 emitters in absolute terms

Source: IEA, 2019

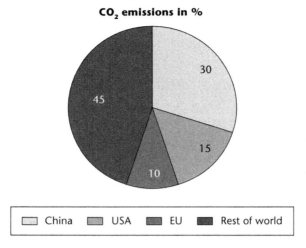

FIGURE 7.2 The world's top CO_2 emitters in percentages: China, USA, EU and the rest of the world

Source: IEA, 2019

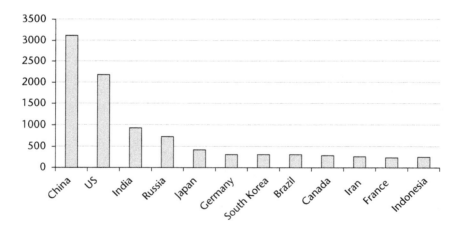

FIGURE 7.3 The world's top energy consumers in absolute terms

Source: IEA, 2019

all countries and not only the developed countries. While future emissions will be very high, the current per capita contribution of emerging economies to climate change tends to be lower than the per capita contribution of industrialised countries, particularly in North America and Australia. There is, however, the exception of China and South Africa, which have similar average emissions per capita to the European Union. Per capita emissions are extremely high in some OPEC (Organization of the Petroleum Exporting Countries) countries, such as Qatar, United

Energy consumption in %

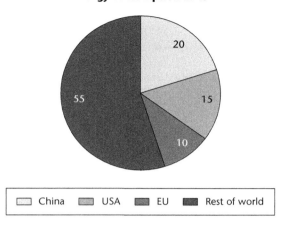

FIGURE 7.4 The world's top energy users in percentages: China, USA, EU and the rest of the world

Source: IEA, 2019

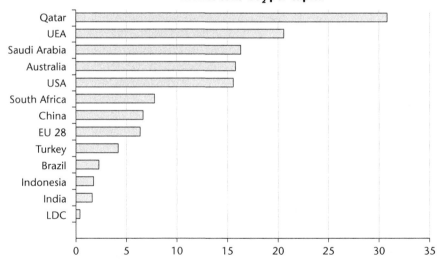

FIGURE 7.5 CO_2 emissions per capita per year worldwide

Source: IEA, 2019

Arab Emirates and Saudi Arabia. This is in stark contrast to the marginal per capita emissions in the Least Developed Countries (LDCs), which are extremely low.

This is shown in Figure 7.5.

The same trend can be observed for energy use: per capita energy use and especially per capita electricity use remain low in developing countries, even in many emerging economies, compared with industrialised countries, despite increasing consumption. For example, the electric power consumption per capita in India was about 20 times lower and in China about four times lower than the per capita consumption in the USA in 2016 (World Bank, 2019). Some scholars are therefore suggesting that personal carbon budgets would be a useful and equitable way to mitigate global climate change (e.g. Fawcett, 2010).

Business responsibility

Most of the discussions on climate change mitigation are happening at the national level. Other initiatives exist, such as in the USA, where national climate and energy policies are lagging behind, yet climate action is happening at the local level, such as in several federal states (e.g. California) and in several cities.

More radical suggestions include a discussion of the responsibilities that firms bear for global climate change. Heede (2014) analysed the historic responsibility of firms for using up emission space between 1854 and 2010. The top five business emitters were reported to be Chevron (USA), ExxonMobil (USA), Saudi Aramco (Saudi Arabia), British Petroleum BP (UK) and Gazprom (Russia). Unsurprisingly, all of these firms are major oil and gas firms (Heede, 2014). A recent study by the European Commission (2019) analysed the major carbon dioxide emitters in the EU and found that the top nine spots were taken by major coal-powered plants running on lignite, one based in Poland, one based in Bulgaria and seven based in Germany. These power plants are operated by PGE (Poland), RWE (Germany), EPH (Czech Republic), Vattenfall (Sweden), AES (USA) and EPH (Czech Republic) (European Commission, 2019; The Guardian, 2019). Interestingly, several coal-fired power plants were sold off by Vattenfall to EPH a few years ago. Spot number 10 was filled by the airline Ryanair, who is the only top corporate emitter in the EU that is not an energy business (European Commission, 2019; Transport and Environment, 2019). International aviation and international shipping are considered bunker fuels, which means they are often excluded from national energy and emissions statistics, as it is hard to pinpoint exactly where the emissions should be allocated to: Is it the country from which the plane or the vessel departs? Or the country in which the plane or vessel arrives? Or the country under which the airline or shipping company is registered? This is especially difficult for international shipping as the vessels are often operating under a different flag than the country in which the vessel was built or where it left the port, and the cargo often comes from many different countries. Hence, all of these allocation approaches are raising ethical issues, be it at the national state level, at the business level or at the personal (per capita) level.

Common but differentiated responsibilities and climate justice

The International Declaration of Human Rights of 1948 advocates global equity. According to these principles of equity, equality, fairness and justice, it should be the incontestable right of developing countries and every person who lives there to develop and to meet their growing needs, even though consumption levels and population levels increase rapidly in many developing countries. It also has to be considered that a significant share of the industry in many developing countries is manufacturing low-cost goods, which are exported and consumed in developed parts of the world such as North America, Europe and Australia. The effects of this embedded energy, like the environmental damage and the emissions from this production, largely occur in developing countries, even though the goods are mostly not consumed within the country of production. About a quarter of China's total carbon emissions are estimated to be due to net export, mainly to the USA and the EU (Wang & Watson, 2007). The craving for low-cost consumption in industrialised countries may be one of the reasons why total GHG emissions and total energy consumption are growing in some developing countries. In return, some agricultural and food products are produced in industrialised countries and are exported to developing countries (and vice versa).

The Kyoto Protocol

In line with these principles of equity, equality, fairness and justice, it has to be acknowledged that even though climate change is a global issue, there are 'common but differentiated responsibilities' for different countries as stated in the first climate treaty that pre-dates the Paris Agreement, namely the Kyoto Protocol (UNFCCC, 1998: 9). It was ongoing between 2005 and 2020 (first and second commitment period). Schreurs and Tiberghien (2007) argue that the EU, particularly Sweden, the Netherlands, Denmark, Germany and the UK played a key role in the establishment, ratification and implementation of the Kyoto Protocol. Though many developing countries (non-Annex I countries) approved, accepted or ratified the United Nations Framework Convention on Climate Change (UNFCCC) and the Kyoto Protocol, non-Annex I countries do not have binding commitments for emission reductions. However, there are incentives for non-Annex I countries to sign up to the UNFCCC and the Kyoto Protocol. These incentives are partly financial and partly in terms of technology access, such as through the Clean Development Mechanism (CDM) (Ockwell et al., 2010).

From Kyoto to Paris – roadmap to a new global climate treaty

In recent years, the situation has changed as emerging economies such as Brazil, China, India, Indonesia, Mexico, Turkey and South Africa have also become emergent emitters. The emissions of emerging economies will increase even more in the future. It is therefore suggested by many that developing countries should have binding targets to avoid ineffective global climate deals and freeriding. The willingness of non-Annex I countries to take on emission reduction targets is, however, linked to their mitigation capabilities, which in turn is linked to their access to modern mitigation technology and financial support, as well as their economic performance and their energy endowments (Rong, 2010). Their willingness is also linked to their request for deep cuts in emissions in industrialised (Annex I) countries (Den Elzen et al., 2008; Rong, 2010). Den Elzen et al. (2008) suggest that to avoid dangerous climate change, non-Annex I countries need to take on binding, quantifiable emission targets by 2030 at the latest. One needs to keep in mind that non-Annex I countries are developing countries that did not cause the climate problem in the first place. Some of them, such as China and India, are already major emitters, others, such as low-income countries, have very low, marginal emissions. As a result of the international climate change negotiations in Copenhagen and Cancun at the Conference of the Parties COP 15 and COP 16, 139 countries acknowledged the need for a new post-2012 climate change agreement which includes not only the Annex I countries, but also the major emitters such as China, India and the USA (the USA has not ratified the Kyoto Protocol). Nevertheless, the Copenhagen Accord and the Cancun Agreements did not include any binding emission reduction targets, but loosely refer to a 'strong political will to combat climate change' (UNFCCC, 2009, 2010). From late 2009 onwards, about 45 developing countries submitted their Nationally Appropriate Mitigation Actions (NAMAs), which are proposed actions that could reduce emissions in non-Annex I countries and are in line with national development and economic priorities. This included emerging emitters such as China and India as well as least developed countries such as Ethiopia that have marginal emissions but ambitious plans for low carbon development as a way to reduce poverty and stimulate economic growth (Urban et al., 2013). At COP 17 in Durban in 2011, it was decided to delay a global future climate agreement under the United Nations Framework Convention on Climate Change (UNFCCC) regime until 2015, with implementation from 2020. In addition, a few major emitters such as Russia, Japan, Canada, Belarus, Ukraine and New Zealand withdrew from the second commitment period of the Kyoto Protocol. This leaves only a small number of Annex I countries, mostly the countries of the European Union, committed to the second round of the Kyoto Protocol between 2013 and 2020. Nevertheless, these countries only account for 15% of global emissions (IEA, 2019). COP 18 in Doha and COP 19 in Warsaw reinforced the message that the emerging emitters of China and India will only be obliged to reduce their emissions after 2020, when (or if) a new global climate deal will be reached.

The Paris Agreement and climate justice

A milestone was reached in 2015 in Paris at COP 21, when a new international climate treaty, the Paris Agreement, was negotiated to start in 2020. The aim is to limit global atmospheric temperature increase to below 2 degrees Celsius. The Paris Agreement was signed by 195 countries. It came into effect in November 2016 when 55 UNFCCC parties, which together accounted for 55% of the global emissions, signed the agreement. The Paris Agreement will first start operating in the period 2020 to 2030 and will require emission reductions by all signatories, both developed and developing countries. It will also enable climate change mitigation, as well as aim for climate change adaptation, enable climate finance and facilitate access to climate-relevant technologies. Schreurs (2016) suggests that the Paris Agreement only came into being once China, the US and the EU had indicated major plans to reduce their domestic emissions.

Climate justice is a concept that emerged throughout the UNFCCC climate negotiations, and the Paris Agreement aims to embed its principles. Tanner and Harvey (2013) explain that:

> Social justice is most commonly used to refer to equity in society, which refers to the state, quality or ideal of being just, impartial, and fair. This is underpinned by morality; judgements about right and wrong that people hold and act upon in their daily lives. It is informed by our understanding of ethics, which concerns the systematic evaluation of such beliefs.
>
> *Tanner and Harvey (2013: 56)*

There is a distinction between social justice within procedures and outcomes. Social justice within climate change and energy procedures, also referred to as procedural social justice or equity, deals with the position of people and groups in climate and energy processes and decision-making. It related to issues of power and participation, as well as competing interests and ideas (Tanner & Harvey, 2013). Social justice with climate change and energy outcomes, also referred to as distributive social justice or equity, deals with the distribution of costs and benefits of energy and climate change among people and groups. It relates to people's entitlements as well as the need for others to recognise these entitlements. Within global climate change debates, notions of social justice have often been about urging wealthier countries to act in light of the distributive social injustice of climate causes and effects (e.g. the wealthy countries being responsible for the bulk of climate change, but the impacts being mainly felt in the poor countries) (Tanner & Harvey, 2013). With regard to equity and justice in the climate change debate, the very fact that developing countries, including emerging emitters, develop should not be contestable. However, the way in which developing countries develop may be debated. There are a number of options to enable more sustainable development, or even low carbon development, and transitions to renewable and low carbon energy are one of these options. Sustainable development is defined by the Brundtland Commission as development 'meeting the needs of the present without compromising the ability of future generations to meet their own needs' (UN, 1987: 1). This is the basis for the Paris Agreement and the aim to meet the 2 degree target.

Energy and climate policy of major emitters – China, USA, India, EU

This part discusses the energy and climate policy of major emitters as part of the Nationally Determined Contributions (NDCs) for achieving the aims of the Paris Agreement. All signatories of the Paris Agreement were required to submit their

NDCs. This system ensures that each country decides themselves which mitigation and adaptation actions are appropriate and realistically achievable, based on its national priorities and capabilities. The basic idea behind this approach is that each country is able to pursue climate action that is in line with its national development agenda, hence enabling climate action while safeguarding wider socioeconomic integrity. However, this means that the process is policy based rather than scientifically based. Climate science is very clear that the Paris Agreement's goal of limiting global atmospheric temperature increases to 1.5–2 degrees is crucial to avoid dangerous climate change (IPCC, 2014, 2018). Yet, at the moment the combined NDCs of the signatory parties are likely to lead to a warming of about 3 degrees or more. Hence, more stringent climate action is needed. As discussed above, China and the USA are currently the major emitters, contributing to 30% and 15% of global CO_2 emissions, respectively (IEA, 2019). It can therefore be argued that it is globally important what happens in China's and the USA's domestic energy and climate policies. Hence, what China and the USA do at the domestic level will have repercussions worldwide in terms of climate change. In the next section, we will therefore review the NDCs of China and the USA, as well as those of the other major emitters India and the EU.

China's energy and climate policies in a nutshell

At present, China accounts for about 20% of global energy consumption, about 30% of global CO_2 emissions and has a high dependency on coal – making up nearly 70% of the electricity generation. Emissions from the country's transport sector are also growing (IEA, 2019). On the other hand, China is also the world's largest investor and installer of renewable energy including hydropower, wind energy and solar PV. The Chinese per capita emissions are about 6.5 t CO_2/person – similar to EU level (IEA, 2019). China also has a high share of embedded carbon emissions due to trade, as many products are being produced in China but consumed elsewhere such as in the USA and in Europe. China has relatively low historic emissions, but high future emissions also play a role. The government has placed a strong focus on climate change and energy issues in recent years, for several reasons: It sees climate change mitigation as an opportunity to restructure the economy, increase national competitiveness, create employment and achieve early mover advantages (e.g. in battery technology). Other reasons are to help the country to adapt to climate change as China has been severely affected by droughts, floods, storms and other climate-induced impacts. Finally, air pollution is a domestic driver for reducing the reliance on fossil-fuel-based energy generation, industry and transport systems, hence leading to co-benefits with climate change mitigation.

China's Nationally Determined Contribution NDC

China's NDC aims to peak 'CO_2 emissions around 2030 and making best efforts to peak early, to reduce CO_2 emissions per unit of GDP by 60% to 65% by 2030 compared to 2005' (UNFCCC, 2015a: 5). Please note this goal is related to carbon intensity reduction; it does not require an absolute cut in emissions. Other goals are 'to increase the share of non-fossil fuels in primary energy consumption to around 20% by 2030' and 'to increase the forest stock volume by around 4.5 billion cubic meters by 2030 compared to 2005' (UNFCCC, 2015a: 5).

China aims to achieve these NDC goals by increasing the promotion of a wide range of technologies and approaches such as renewable energy like wind, solar, bioenergy and geothermal energy; 'clean coal' technologies; natural gas; nuclear energy; building efficiency; smart grids; district heating; improved waste management; low carbon transport systems; 'low carbon lifestyles' and enhancing carbon sinks like forests (UNFCCC, 2015a).

The USA's energy and climate policies in a nutshell

The USA accounts for about 15% of global energy consumption and about 15% of global CO_2 emissions (IEA, 2019). The country has a heavy reliance on fossil fuels – oil for transport and coal and gas for electricity (65% of the total electricity) (IEA, 2019). In addition, the US per capita emissions are high, at about 15 t CO_2/person; this is more than double that of China and the EU (IEA, 2019). Also, historic emissions are high (IEA, 2019), and the USA has been outsourcing and offshoring some emissions to Asia and other parts of the world through relocating some of its business operations overseas. Trade plays an important role here as the products that are being produced in China and other parts of the world are partly being consumed in the USA; hence, there is a mismatch in the allocation of emissions between the producers and the consumers. US climate and energy policy has been challenged internally by Trump's threats to leave the Paris Agreement, which was negotiated under Obama. Due to the lack of political commitment and in the absence of any advanced national climate and energy policy, individual states and cities are driving forward action on climate and energy targets.

USA's NDC

The USA aims to 'reduc[e] greenhouse gas emissions by 26%–28% by 2025 compared to 2005 and to make best efforts to reduce its emissions by 28%' (UNFCCC, 2015b: 1). The NDC is referring to existing legislations such as the Clean Air Act, the Energy Policy Act and the Energy Independence and Security Act, as well as referring to existing building codes and vehicle standards. No specific action is outlined, neither at the sectoral level nor in relation to specific technologies. The baseline year of 2005 is similar to China, yet different to the EU, which has chosen 1990 as the baseline. Any climate targets that set 2005 as a baseline are likely to be less ambitious than those that use a 1990 baseline due to the economic growth that happened during this time, which has also required increased energy supply and resulted in increased emissions (IEA, 2019).

India's energy and climate policies in a nutshell

India accounts for about 5% of global energy consumption and about 5% of global CO_2 emissions (IEA, 2019). India has a heavy reliance on fossil fuels – oil for transportation, while about 75% of the electricity is generated from coal (IEA, 2019). The country is expanding both its fossil fuel capacity and its renewable energy capacity. Indian per capita emissions are very low at 1.6 t CO_2/person. This is almost 10 times lower than the per capita emissions in the US (IEA, 2019). Embedded carbon emissions due to trade play a role, such as for the textile industry that manufactures clothes for OECD countries. India has low historic emissions, but high future emissions also play a role.

India's NDC

India aims to 'reducing the emissions intensity of its GDP by 20–25%, over 2005 levels, by 2020' (UNFCCC, 2015c: 8). Again, the measurements here are carbon intensity and a different end year was chosen. India has a wide range of policies and plans in place to achieve these targets. It has targets for specific energy technologies, such as wind energy, solar, biomass, hydropower, nuclear and 'clean coal', as well as having a smart grid mission in place, plans for enhancing energy efficiency, including in the industrial sector and for improving energy conservation in buildings. Other targets are related to climate smart cities, improved waste management, increasing urban public transport and promoting electric vehicles. Other mitigation actions include planned afforestation.

India also aims to provide universal electricity access to all its residents in the near future (UNFCCC, 2015c).

The EU's energy and climate policies in a nutshell

Twenty-eight countries[1] of the EU jointly account for about 10% of global energy consumption and about 10% of global CO_2 emissions (IEA, 2019). About 40% of the electricity comes from fossil fuels and about 35% from renewables (IEA, 2019). The EU as a whole has in fact declining CO_2 emissions, which is very positive. Per capita emissions in the EU are in average about 6 t CO_2/person which is similar to China (IEA, 2019). Some countries are leaders in low carbon energy transitions (e.g. Sweden, Iceland, Norway, Denmark, France, Switzerland, partly Germany). Despite these positive developments, there are several negatives too: Historic emissions are high (IEA, 2019), and the EU has outsourced and offshored some of its emissions to Asia due to trade. There are a few exceptions, such as Germany which has long been known as an economy that exports more goods than it imports, particularly high-end exports such as cars and other vehicles, machinery and mechanical devices (including washing machines and other white goods), as well as electronics. Also, some countries remain unambitious about climate policy, such as Poland and Estonia, which rely heavily on coal for electricity generation (as does Germany partly) as well as Malta and Cyprus where nearly all electricity comes from oil (IEA, 2019).

EU's Nationally Determined Contribution NDC

The target of the EU is to achieve a 40% cut in greenhouse gas emissions compared to 1990 levels by 2030 and to reduce its emissions by 80%–95% by 2050 compared to 1990 (UNFCCC, 2015d: 1 and 3). Other targets are to achieve at least a 27% share of renewable energy consumption and an improvement in energy efficiency at EU level of 27%–30% by 2030. These targets cover a wide range of climate actions in the sectors of energy, industry, transport, waste management, land use and forestry, etc. In addition to these EU-wide targets, every EU member state has their own national targets that are in line with their national needs and priorities. Some countries have an especially ambitious climate policy, such as Sweden, for example. The Swedish government aims to achieve 100% renewable electricity by 2040 and for the country to become carbon neutral by 2045 (Energimyndigheten, 2019). Today electricity generation in Sweden is almost completely fossil-free – Sweden relies to 40% on hydropower, 40% on nuclear power, 10% on wind energy and 9% on other renewables such as modern biomass (IEA, 2019). The country has very special geographic conditions for hydropower and its large forest base means that major cities are predominantly heated by large-scale district heating systems based on biomass. There are initiatives in Sweden's industry for fossil-free steel production (e.g. HYBRIT) and the transformation of the transport sector is on-going with an increasing share of electric vehicles (EVs), as well as hybrid and biofuel vehicles, e.g. as part of public transport fleets.

Exercises

1   Look at the Nationally Determined Contributions (NDCs) of your country of choice: https://www4.unfccc.int/sites/NDCStaging/Pages/All.aspx

   What energy and climate actions have been proposed? How does this relate to the current socioeconomic, political and technological status of your country of choice? How does this compare to other countries in the same income group?

2   Imagine you are a policymaker who has to explain your country's energy and climate strategy to the parliament. How would you describe your countries' contribution to climate change from (a) the historical perspective, (b) the current perspective and (c) the future perspective? (d) How would you explain the role of business (e.g. energy utilities, manufacturing industry, transport companies etc.) and (e) the role of individuals (e.g. personal carbon budgets)? (f) What role would trade play for the country of your choice? (g) What role does energy justice and climate justice play here? You may wish to prepare a short speech or a short PowerPoint presentation.

Note

1 At the time this chapter was written (August 2019), the EU still had 28 member states. Once the UK leaves the EU because of Brexit, the EU will have 27 member states.

References

Climate Analytics (2015) *Historical Responsibility for Climate Change – From Countries Emissions to Contribution to Temperate Increases*. Available from: https://www.climateanalytics.org/media/historical_responsibility_report_nov_2015.pdf

Den Elzen, M., Höhne, N. & Moltmann, S. (2008) The Triptych approach revisited: A staged sectoral approach for climate mitigation. *Energy Policy*, 36 (3), 1107–1124.

Energimyndigheten. (2019) Sweden's energy and climate targets (in Swedish). Available from: http://www.energimyndigheten.se/klimat--miljo/sveriges-energi--och-klimatmal/

European Commission. (2019) *Verified Emissions from 2018 – Data*. Available from: https://ec.europa.eu/clima/policies/ets/registry_en#tab-0-1

Fawcett, T. (2010) Personal carbon trading: A policy ahead of its time? *Energy Policy*, 38, 6868–6876.

Heede, R. (2014) Tracing anthropogenic carbon dioxide and methane emissions to fossil fuel and cement producers, 1854–2010. *Climatic Change*, 122 (1–2), 229–241.

IEA. (2019) *Energy and Emissions Data*. Available from: www.iea.org/statistics

IPCC. (2014) *Climate Change 2014*. Synthesis Report. Available from: https://www.ipcc.ch/report/ar4/syr/full-report/

IPCC. (2018) *Global Warming of 1.5 Degrees*. Available from: https://www.ipcc.ch/sr15/

Ockwell, D., Mallett, A., Haum, R. & Watson, J. (2010) Intellectual property rights and low carbon technology transfer: The two polarities of diffusion and development. *Global Environmental Change*, 20, 729–738.

Rong, F. (2010) Understanding developing country stances on post-2012 climate change negotiations: Comparative analysis of Brazil, China, India, Mexico and South Africa. *Energy Policy*, 38 (8), 4582–4591.

Schreurs, M.A. (2016) The Paris climate agreement and the three largest emitters: China, the United States, and the European Union. *Politics and Governance*, 4 (3), 219–223.

Schreurs, M.A. & Tiberghien, Y. (2007) Multi-level reinforcement: Explaining European Union leadership in climate change mitigation. *Global Environmental Politics*, 7 (4), 19–46.

Tanner, T. & Harvey, B. (2013) Social justice and low carbon development. In: Urban, F. & Nordensvärd, J. (Eds.) *Low Carbon Development: Key Issues*. Oxon, Earthscan, Routledge. pp. 55–65.

The Guardian. (2019) *'Ryanair Is the New Coal': Airline Enters EU's Top 10 Emitters List*. Available from: https://www.theguardian.com/business/2019/apr/01/ryanair-new-coal-airline-enters-eu-top-10-emitters-list

Transport and Environment. (2019) *State of the Aviation ETS*. Available from: https://www.transportenvironment.org/state-aviation-ets

UN. (1987) *Report of the World Commission on Environment and Development: Our Common Future – The Brundtland Report*. Available from: https://sustainabledevelopment.un.org/content/documents/5987our-common-future.pdf

UNFCCC. (1998) *Kyoto Protocol*. Available from: https://unfccc.int/resource/docs/convkp/kpeng.pdf

UNFCCC. (2009) *Copenhagen Accord*. Available from: https://unfccc.int/resource/docs/2009/cop15/eng/11a01.pdf

UNFCCC. (2010) *Cancun Agreements*. Available from: https://unfccc.int/process/conferences/pastconferences/cancun-climate-change-conference-november-2010/statements-and-resources/Agreements

UNFCCC. (2015a) *China's Nationally Determined Contribution*. Available from: https://www4.unfccc.int/sites/ndcstaging/PublishedDocuments/China%20First/China%27s%20First%20NDC%20Submission.pdf

UNFCCC. (2015b) *USA's Nationally Determined Contribution*. Available from: https://www4.unfccc.int/sites/ndcstaging/PublishedDocuments/United%20States%20of%20America%20First/U.S.A.%20First%20NDC%20Submission.pdf

UNFCCC. (2015c) *India's Nationally Determined Contribution*. Available from: https://www4.unfccc.int/sites/ndcstaging/PublishedDocuments/India%20First/INDIA%20INDC%20TO%20UNFCCC.pdf

UNFCCC. (2015d) *EU's Nationally Determined Contribution*. Available from: https://www4.unfccc.int/sites/ndcstaging/PublishedDocuments/European%20Union%20First/LV-03-06-EU%20INDC.pdf

Urban, F., Ting, M.B. & Lakew, H. (2013) Poverty reduction and economic growth in a carbon constrained world: The case of Ethiopia. In Urban, F. & Nordensvärd, J. (Eds.), *Low Carbon Development: Key Issues*. Earthscan, Routledge, Oxon, pp. 228–239.

Wang, T. & Watson, J. (2007) *Who Owns China's Carbon Emissions?* Tyndall Centre for Climate Change Research, Brighton. Briefing Note No. 23.

World Bank. (2019) *Open Data*. Available from: https://data.worldbank.org/

World Resources Institute (WRI). (2005) *World Greenhouse Gas Emissions in 2005*. WRI Working Paper. Available from: www.wri.org/publication/world-greenhousegas-emissions-2005

8

THE HEALTH IMPLICATIONS OF ENERGY USE

Health and well-being implications of traditional biomass use

As about 1 billion people worldwide do not have access to electricity and 2.7 billion people rely on traditional biomass for basic needs such as cooking and heating (IEA, 2019), a large share of the global population still relies on traditional biomass, which causes indoor air pollution when used for cooking and heating.

Health implications of indoor air pollution

The most important health implication of traditional biomass use is linked to indoor air pollution. Indoor air pollution is caused by the combustion of traditional solid fuels, such as fuelwood, charcoal, dung and agricultural residues. This causes the release of smoke and toxic substances such as sulfur oxides (SO_x), nitrogen oxides (NO_x), volatile organic compounds (VOCs), particulate matter (PM) and carbon monoxide (CO).

> A condition of 'air pollution' may be defined as a situation in which substances that result from anthropogenic activities are present at concentrations sufficiently high above their normal ambient levels to produce a measurable effect on humans, animals, vegetation, or materials.
>
> *Seinfeld and Pandis (2006: 21)*

Indoor air pollution is associated with a series of negative health impacts including pneumonia, chronic obstructive respiratory disease, lung cancer and adverse pregnancy outcomes (WHO, 2000, 2005b, 2006).

According to the WHO (2019), about 3.8 million people – mostly women and children – are likely to die every year because of exposure to indoor air pollution

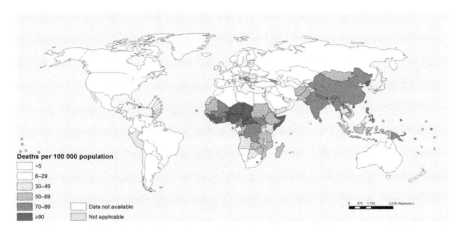

Deaths per 100 000 population
- <5
- 6–29
- 30–49
- 50–69
- 70–89
- ≥90
- Data not available
- Not applicable

FIGURE 8.1 Deaths attributable to household indoor air pollution in 2016, measured as deaths per 100,000 population

Source: WHO, 2019

from traditional biofuels. This is more people than die each year from malaria or tuberculosis. About 4.2 million people die prematurely because of outdoor air pollution The WHO also estimates that about 90% of the people world-wide live in areas where air quality exceeds WHO guideline limits (WHO, 2019). Figure 8.1 shows the unequal distribution of deaths attributable to household indoor air pollution, which is linked with energy poverty hotspots such as in sub-Saharan Africa and developing Asia. Dark fields show the highest number of deaths.

A study from the University of California Berkeley argues that exposure to indoor air pollution may be linked to lower neurodevelopmental performance and potentially even lower IQs for children whose mothers were exposed to indoor air pollution when pregnant (Yang, 2011). Women and children are more likely to be affected by indoor air pollution and its adverse effects as they tend to stay close to the local hearth and are responsible for cooking and domestic activities. Introducing modern energy sources, such as electricity, biogas, natural gas or solar cookers, to replace cooking with traditional biofuels is likely to increase the health of the population. Improved cook stoves can help to increase energy efficiency and thereby reduce the amount of fuel used for cooking and reduce indoor air pollution.

Health implications of fuelwood collection

Another social impact associated with fuelwood is due to fuelwood collection. Fuelwood collection takes up significant amounts of time every day and is mostly done by women and children, especially girls. The social implication is that women and children spend time collecting fuel rather than spending time in education, productive/income-generating activities or household and family activities. Health impacts include strain on the body from labour-intensive fuelwood collection. In some cases, injuries and body deformations can arise due to the burden of transporting fuelwood (WHO, 2000, 2005b, 2006).

A method to calculate the burden of disease from mortality and morbidity is using a disability adjusted life year (DALY). One DALY is considered by WHO as one lost year of 'healthy' life. Summing up all the DALYs across the population is like measuring the gap between current health status and an ideal health situation where the entire population lives to an advanced age, free of disease and disability (WHO, 2016). Currently, the Gold Standard is developing a methodology to calculate health benefits using averted disability adjusted life years (ADALYs) as the impact metric. This is based on a methodology developed by the World Bank and the University of Berkeley and will contribute to estimating the finance payment mechanisms needed to drive initiatives to improve lives (Gold Standard, 2016).

The health implications of fossil fuel use

Health and well-being implications of fossil fuel use

After discussing the health implications from traditional solid fuels, such as fuelwood and dung, we will discuss the health implications of fossil fuel use. There are two major health impacts of fossil fuels: outdoor air pollution and climate change.

In contrast to indoor air pollution, outdoor air pollution happens outside of the home. This type of pollution is often found in cities and urban, industrialised areas. While indoor air pollution affects mainly the users of energy, outdoor air pollution affects everyone breathing the outdoor air, and therefore can affect people who have not caused the pollution. One of the problems with long-range and transboundary air pollution from fossil fuel combustion of coal, oil and natural gas products is that the pollution can be caused far away from the source area, for example, pollution sources in urban areas can cause adverse effects in rural areas and even cross boundaries affecting other countries. A striking example is the case of the German Democratic Republic (GDR). Up until 1989, pollution from the fossil fuel industries in the GDR was blown with the wind over the border into Germany and affected people in places such as Bavaria. Outdoor air pollution is a problem that gained international importance due to industrialisation. It results from the combustion of fossil fuels, particularly coal, which is high in sulfur, thereby releasing sulfur dioxide emissions and smoke when burned. Fossil fuels are combusted for heating, industrial purposes and transport. Fossil fuel combustion from transport in particular releases carbon monoxide. The combination of smoke and fog is called smog and is common both in the winter in cold climates and in the summer in warm climates. Particulate matter (PM) is especially high in smog. Three cities that are famous for smog are London, particularly seen from a historical perspective as air pollution has become better in recent decades; Los Angeles, which suffers from smog on warm, sunny days; and Beijing, which suffers from smog both on cold winter days and warm, sunny days.

The smog on warm, sunny days is caused by tropospheric ozone (O3), ozone that is present in the lowest layer of the Earth's atmosphere, called the troposphere. The ozone undergoes a photochemical reaction in the atmosphere (Smithson et al., 2008). The smog on cold days is especially pronounced in stagnant air with little air circulation and in temperature inversions. Usually, warm air is located near the ground and the air can rise and carry away pollutants. Temperature inversions occur when cold air is trapped near the ground by a layer of warmer air, which means that toxic substances, such as carbon monoxide, stay trapped close to the ground. The use of more modern fuels, such as electricity and natural gas, has reduced sulfur dioxide emissions, and policies to reduce air pollution have also played an important role.

While historically the 'policy of high chimneys' for polluting industries and a move towards modern energy enabled a reduction in air pollution in many OECD countries, such as the UK, the recent problem in air pollution in many cities, such as Los Angeles and Beijing, is mainly caused by excessive traffic. This situation is being made worse by increasing private car ownership in many Asian cities, such as Beijing. Beijing has introduced several policies to restrict personal vehicle use, such as restricting vehicle use on certain week days based on the registration number, partly as a result of policies introduced for the 2008 Olympics (Hao et al., 2011; Cai & Xie, 2011).

Apart from industrial emissions of sulfur dioxide that cause air pollution, transportation leads to emissions of oxides of nitrogen, particles, hydrocarbons and carbon monoxide, which may significantly degrade urban air quality. This leads to a range of negative health implications. At relatively high concentrations, nitrogen dioxide causes inflammation of the respiratory tract, affects lung function and exacerbates the response to allergens in sensitised individuals. A further consequence of nitrogen dioxide (and sulfur dioxide) emissions is the formation of acid precipitation (commonly known as acid rain), which leads to various forms of ecological damage, such as the acidification of lakes and other water bodies. This in turn can have health implications due to links to drinking water, the food chain and recreational activities (e.g. bathing in acidified lakes). Particles (also termed 'particulate matter', PM or 'particulates') have a range of health implications: They affect the cardiovascular and respiratory systems, exacerbate asthma and cause a direct increase in mortality. The health effects of particles may be especially severe in people with pre-existing heart and lung diseases. There are various other pollutants, for example 'hydrocarbons'. This covers a diverse group of organic chemicals including benzene, polyaromatic hydrocarbons, kerosene, diesel and de-icing compounds (such as ethylene glycol), some of which are known genotoxic carcinogens. Also, emissions of carbon monoxide can be toxic to humans due to the formation of carboxyhaemoglobin, which reduces the oxygenation of blood and tissues and which poses a particular risk to individuals with pre-existing cardiovascular or respiratory diseases. The secondary pollutant ozone causes irritation to the eyes, damage to the respiratory tract and also triggers inflammatory responses. In addition to these key substances of concern, a vast range of other air pollutants has been identified (Seinfeld & Pandis, 2006). It is worth emphasising that carbon dioxide – while being an important greenhouse gas (GHG) that contributes to climate change – is not regarded as a pollutant that degrades air quality or that directly affects human health.

Overall, in Europe, poor air quality is responsible for severe impacts on human health, including around 370,000 early deaths in 2000; air quality concerns are also acute in other parts of the world (Smithson et al., 2008). Morgenstern et al. (2008) found a close correlation between children suffering from asthmatic bronchitis and allergies, such as hay fever and eczema, and their exposure to traffic-related air pollution.

These health impacts do not only represent a high burden for those affected, but also pose high financial burdens to individuals, taxpayers, states and healthcare institutions.

Figure 8.2 shows the estimated deaths from urban outdoor air pollution in 2016, based on WHO estimates.

There is a wide range of well-being and social implications of fossil fuel use, which are partly related to the environmental impacts of fossil fuel use. This relates, for example, to the conditions under which coal mining is conducted in many parts of the world. While work in the oil and gas industry generally pays well and minimum safety standards are often adhered to (with some exceptions), coal mining is still a badly paid and risky job in many parts of the world (far less so in OECD countries). There are also adverse health effects, as coal miners often suffer from pneumoconiosis, a lung infection caused by long-term inhalation of dust. In many coal mines in the developing world, such as in China and Latin America, the safety standards are poor and people work under very harsh conditions. At the same time,

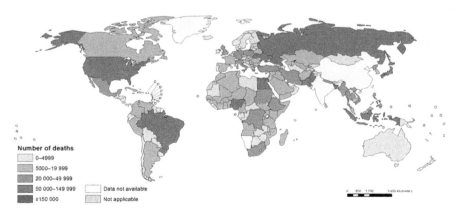

FIGURE 8.2 Deaths from outdoor air pollution

Data source: WHO, 2019

wages are often low, labour rights tend to be limited and deadly accidents occur from time to time.

In some areas, such as the coal-producing Chinese province Shanxi, coal production affects almost every aspect of life and has considerable impacts on the environment, health, employment and the social life of its citizens. Linfen, a city of 3 million inhabitants in Shanxi province, has been named as one of the world's most polluted cities due to the pollution from coal mining, fossil fuel-driven industries and transport. In some places in Shanxi, the houses, cars, trees, grass and everything else is covered with a layer of soot (observations of the author in 2007). The short report below indicates the intensity of the problem:

Number of people potentially affected: 3 000 000
Type of pollutant: Coal and particulates
Source of pollution: Automobile and industrial emissions

This soot-blackened city in China's inland Shanxi province makes Dickensian London look as pristine as a nature park. Shanxi is the heart of China's coal belt, and the hills around Linfen are dotted with mines, legal and illegal, and the air is filled with burning coal. Don't bother hanging your laundry – it'll turn black before it dries. China's State Environmental Protection Agency says that Linfen has the worst air in the country, which is saying something, considering that the World Bank has reported that 16 of the 20 most polluted cities in the world are Chinese.

Walsh (2007)

Newer reports, however, claim that the Chinese government employed substantial efforts to improve the pollution in the city and that is has become a much cleaner and better place to live in the last few years (North, 2012). Meanwhile the situation has worsened in other parts of the world, most notably in India. In 2018, the list of the world's most polluted cities contained more than Indian cities amongst the global top 30, including Delhi, Kolkata and popular tourist destinations like Jodhpur, Varanasi and Agra (WHO, 2019).

Health implications of climate change

The Intergovernmental Panel on Climate Change (IPCC) estimates that about 70% of all GHG emissions worldwide come from energy-related activities. This is mainly from fossil fuel combustion for heat supply, electricity generation and transport and includes carbon dioxide, methane and some traces of nitrous oxide (IPCC, 2007, 2013; Smith et al., 2014). It is well documented that these emissions contribute to global climate change. Energy use has potentially significant climate impacts, which are assumed to exceed the impacts from other sources, such as land use and other industrial activities. Fossil fuel use therefore contributes to global climate change.

The IPCC's Fifth Assessment Report published in 2014 reports that climate change leads to a series of health implications. The direct impacts of a changing climate on health are mainly related to three areas: (1) heat- and cold-related impacts, (2) extreme weather events such as floods and storms and (3) ultraviolet radiation. The IPCC also reports the likely increase of so-called ecosystem-mediated impacts of a changing climate on health, such as vector-borne and other infectious diseases, such as malaria, dengue fever and tick-borne diseases, food- and water-borne diseases and deterioration of air quality (Smith et al., 2014). In particular, the IPCC highlights the following expected health implications of climate change:

1   Greater risk of injury, disease, and death due to more intense heat waves and fires [very high confidence]
2   Increased risk of under-nutrition resulting from diminished food production in poor regions [high confidence]
3   Consequences for health of lost work capacity and reduced labor productivity in vulnerable populations [high confidence]
4   Increased risks of food- and water-borne diseases [very high confidence] and vector-borne diseases [medium confidence]
5   Modest improvements in cold-related mortality and morbidity in some areas due to fewer cold extremes [low confidence], geographical shifts in food production, and reduced capacity of disease-carrying vectors due to exceedance of thermal thresholds [medium confidence]. These positive effects will be outweighed, worldwide, by the magnitude and severity of the negative effects of climate change [high confidence].

Smith et al. (2014: 713)

In addition, the GHGs relevant to air pollution, such as nitrous oxide, also contribute to adverse health implications, as we discussed previously in relation to air pollution. The IPCC further highlights the importance of human vulnerability and the social responses to climate change that impact on human health; for example, the displacement of populations because of prolonged drought. Populations that are already affected by climate-related disasters have a particularly high vulnerability to climate change and are prone to its health impacts; for example, due to existing undernutrition in food-insecure areas. The IPCC reports that climate change is likely to exacerbate pre-existing human health problems, existing diseases are likely to extend their range into new areas that are currently unaffected and new conditions (e.g. diseases) may arise (Smith et al., 2014). To reduce human vulnerability to the health impacts of climate change, the IPCC suggests several mitigation measures:

1   Reducing local emissions of health-damaging and climate-altering air pollutants from energy systems, through improved energy efficiency, and a shift to cleaner energy sources [very high confidence]
2   Providing access to reproductive health services (including modern family planning) to improve child and maternal health through birth spacing and

reduce population growth, energy use, and consequent climate-altering air pollutants emissions over time [medium confidence]

3   Shifting consumption away from animal products, especially from ruminant sources, in high-meat consumption societies toward less climate-altering air pollutants-intensive healthy diets [medium confidence]

4   Designing transport systems that promote active transport and reduce use of motorized vehicles, leading to lower emissions of climate-altering air pollutants and better health through improved air quality and greater physical activity [high confidence]

Smith et al. (2014: 714)

Please note that the IPCC mentions these findings (Smith et al., 2014: 713, 714) with a disclaimer about confidence levels. The confidence level describes the amount of certainty (or uncertainty) that is attached to these projections. These findings are by no means prescriptions of how the future will turn out but are scientific findings of what is likely to happen and how certain or uncertain it is.

The health implications of low carbon energy use

Health and well-being implications of renewable energy

While we previously discussed the health and well-being implications of traditional biomass and fossil fuels, this section discusses the health and well-being implications of low carbon energy. We first focus on renewable energy and large hydropower dams, and then discuss the health and safety implications of nuclear energy.

Renewable energy comes from renewable natural resources, such as sunlight, wind, water, tides, geothermal heat and biomass. Unlike fossil fuels and nuclear energy, which are finite and depletable, these energy resources are renewable and non-depletable. Renewable energy has a large global potential. The World Energy Council estimates that the theoretical potential is 370 PWh/year for solar energy, 315 PWh/year for primary biomass, 96 PWh/year for wind energy and 41 PWh/year for hydropower. Nevertheless, the technical and economic potential is lower due to variations in land availability and financial competition with fossil fuels (WEC, 2018). About 25% of the global electricity consumption came from renewable energy in 2016, mainly from hydropower, but also from wind, solar and geothermal energy, as well as from biomass (IEA, 2019). The most widely used and commercialised renewable energy technologies are wind turbines, solar photovoltaic (PV) panels and hydropower technology. We differentiate here between large hydropower dams and small-scale hydropower, which we include in this section on renewables.

Renewable energy technologies have no major adverse health effects attached to them. They do not have any fuels as an input or any toxic waste as an output, so health issues are related to the production of these energy technologies and siting issues (e.g. for wind farms in close proximity to settlements).

In contrast, renewable energy can help to reduce harmful emissions and thereby reduce health effects. Pursuing low carbon development can be an effective way to reduce the negative health impacts arising from climate change and air pollution. Leading health experts, such as Haines (2012), Haines et al. (2010), Markandya et al. (2009) and Rao et al. (2013), suggest that adopting low carbon policies in a range of sectors could have strong health benefits, as well as enabling costs to be cut. This includes for example switching to renewable energy in the power sector, promoting low carbon transport options, introducing lower emissions industrial strategies, etc.

Health and well-being implications of large hydropower dams

Dams are large hydropower plants that hold back and control the natural water flow to generate hydroelectric power. Large dams involve building of a reservoir, a large area of stored water. While hydropower is considered a low carbon or climate-friendly energy source (although there are some GHG emissions from dams, such as methane emissions from the decaying organic material of flooded vegetation in the reservoir), dams can have major impacts on people's lives. First, they can lead to displacement and resettlement of local communities and individuals, and second, they can affect the livelihoods and lifestyles of local communities and individuals, which in turn affect well-being, physical health and mental health of those directly affected.

Displacement, resettlement and impacts on livelihoods and lifestyles

Around the world, dam building has led to the displacement and relocation of millions of people. In some cases, it has been reported that affected people were relocated to areas without appropriate infrastructure that lacked basic amenities, such as sanitary facilities, drinking water, electricity and roads, for example, at Tarbela (Pakistan) and Tucuruí (Brazil) (Asianics Agro-Dev. International (Pvt) Ltd, 2000; La Rovere & Mendes, 2000). Lacking access to safe drinking water resulted in various health impacts, such as a higher prevalence of diseases and risk of dehydration.

In other cases, and most notably in China, people were relocated from rural areas to modern cities where they had improved access to infrastructure and modern amenities, including access to modern healthcare, but lost their rural livelihoods. The Chinese Three Gorges Dam included the flooding of 13 cities, 140 towns and 1350 villages and the flooding of numerous sites of cultural, historic and religious heritage. For the Three Gorges Dam, 1.3 million inhabitants were relocated, many of them from rural areas to the cities (Rissler, 2002; International Rivers, 2008). It has been reported that many subsistence farmers and fishermen were relocated to urban areas, or they received tiny slots of barren land as compensation. As a result, many people were worse off after the resettlement than before due to loss of livelihoods, increased unemployment, decreased income and insufficient land for subsistence farming. Scientific studies suggest that the resettlement process and the loss of cultural and social roots at the Three Gorges Dam resulted in decreased mental health, induced by mental stress, and an increase of depression among the resettled population (Hwang et al., 2007).

Other key social problems regarding relocations include the following, which have adverse effects on well-being and mental health: compensation payments are often too low for a decent living, relocation of the local population often results in loss of livelihoods, such as fisheries or subsistence farming, and compensation payments are not equally distributed which means that some people do not receive any compensation at all or do so only years after their relocation. With regard to the immediate social impacts of large-scale dam construction, the main issues are as follows: inappropriate resettlement, alterations in the lifestyle of people, the lowering of the standard of living, disregard for local people's rights, identity and culture, and taking land and other natural resource tenure away from local people (Yu, 2003). The psychological trauma of resettlement is one of the biggest issues communities have to grapple with. Loss of landholdings, insufficient compensation, loss of access to natural resources like water and land, and cultural costs compounded with resettlement efforts that lower the standard of living of communities are among the immediate impacts (Adams, 2000; Urban et al., 2013). Other social and well-being effects are related to the change of the natural water flow and water availability downstream. This is relevant for people who depend on the river for water and food security, both in proximity to the dam site and farther away.

Large hydropower dams can have impacts on the livelihoods and lifestyles of communities and individuals, either through displacement and resettlement or through changes to their sources of income and their routines and culture. Let us look at two examples.

In the case of the Bakun dam in Borneo, Malaysia, indigenous groups such as those from the ethnic minorities Kenyah, Kayan, Penan, Lahanan and Iban were relocated due to dam building (Majid Cooke et al., 2017). Before the building started, these indigenous people were hunters, gatherers and subsistence farmers, who lived in a pristine rainforest. Some of them were semi-nomadic before the dam-induced resettlement, e.g. the Penan. Their livelihood strategies were not monetised and not modernised. During dam construction, the indigenous people were displaced and resettled away from the dam site in new-built settlement areas. They lost their traditional lands, and the natural environment has been degraded or destroyed to make way for the dam, the reservoir and access roads. A decline in biodiversity has made hunting and gathering nearly impossible, and a rise in commercial cash crop farming in the rainforest (mainly monoculture oil palm plantations) along the access roads has meant that their traditional lifestyle is no longer possible. Instead, the tribes have turned to commercial activities, including working on oil palm plantations for income. Their lives have become commodified and their traditional lifestyles, routines and culture have been altered because of the dam building. Despite these negative impacts, which may have adversely affected the mental health of the local communities, there were positive overall health impacts. This is because the resettled communities now have access to health clinics in their proximity, as well as access to schools in their proximity. This has made it easier for many families and individuals to access medical care and getting a former education. Another big advantage is that pregnant women have access to medical care within about half an hour's drive on adequate roads, rather than being up-river in the rainforest with the nearest clinic several hours away by boat. This has made childbirth considerably safer and also enabled early medical care for babies, young children and their mothers (Majid Cooke et al., 2017).

Another example is the Kamchay hydropower dam in Cambodia, which is built in Bokor National Park. It is reported that the dam and the reservoir have led to the flooding of 2015 ha of protected forest in Bokor National Park (NGO Forum Cambodia, 2013). There are approximately 22,000 people living in the rural area that is directly affected by the dam (NGO Forum Cambodia, 2013). The largest group of people affected by the dam are people who depend on bamboo collection for their livelihoods. They collect bamboo to make baskets to sell at the local market in Kampot town. Before the large hydropower dam was built, the villagers had an agreement with the Cambodian government that they would be allowed to collect bamboo in the 'multiple-use zone' of Bokor National Park. However, the dam has flooded the bamboo forest area that the villagers used. The dam has therefore meant a decline in livelihoods for many bamboo collectors. A smaller bamboo area is available, but it is farther away and is often closed as it belongs to the dam-operating company Sinohydro/PowerChina. Many bamboo collectors do not have other sources of income, many do not own land and have no assets, and many have low literacy rates and cannot easily transfer to more skilled jobs. Fieldwork revealed that many of the villagers are experiencing financial hardship due to the dam and are considering options such as migration to Phnom Penh or Thailand as migrant workers to secure their livelihoods (Siciliano et al., 2015). Health-wise, this has not led to any improvement. Mental and physical health may be affected as villagers are more exposed to loss of income, financial instability and reductions in food security and hence reduced nutritional intake, which is particularly challenging for children and pregnant women. A small group of Cambodians with Chinese ancestry has more options as their children study at a Chinese school with the aim of being translators for the Chinese dam company and other Chinese enterprises in the area.

Many countries have legislation in place stating that changes to livelihoods and lives of affected individuals and communities should be outlined in an environmental and social impact assessment (ESIA), approved before a dam is built with mitigating measures implemented to reduce negative social and environmental impacts. Nevertheless, in reality, this legislation is often not implemented as stringently as it should be (Siciliano et al., 2015). Also, health impacts of the affected communities are usually not been assessed as part of the ESIA.

Health and well-being implications of nuclear power

Nuclear energy technology refers to power plants that use nuclear fission to generate electricity and heat. At a global level, there was about 400 GW installed nuclear capacity and another 120 GW under construction or planned in 2019. The countries with the largest installed capacity of nuclear power are the USA, France, China, Japan (with much reduced capacity due to the Fukushima nuclear accident in 2011), Russia, South Korea, Canada and Ukraine. Other countries with considerable nuclear operating capacity are Germany (all nuclear power plants to be phased out by the end of 2022), the UK, Sweden, Spain, Belgium and Taiwan (WNA, 2019).

Nuclear energy comes at a risk to health and safety. There are concerns about the increased risks of nuclear radiation for nuclear power plant workers and for people living near nuclear power plants, particularly related to increased occurrence of cancers such as leukaemia in these groups. Another implication of nuclear energy is the radioactive waste that is generated by nuclear power plants. Radioactive waste, such as uranium and plutonium, contains carcinogenic substances, such as radioactive strontium, iodine, caesium and plutonium. The half-life (t½) of radioactive substances can be up to several billion years. The half-life of radioactive uranium, for example, is 4.468 billion years for 238U and 703.8 million years for 235U. The half-life of radioactive plutonium is shorter at 24,000 years for 239P and 6,800 million years for 240P (US NRC, 2002). Half-life is the time that it takes for a quantity of a substance to fall to half its value, for example, the time required to reduce the radioactivity of a substance by half. Radioactive waste with a long half-life must therefore be contained and isolated from the environment and from humans for very long periods of time. A common practice is to deposit radioactive waste in geological formations, such as salt stocks. Nevertheless, the waste will have to be deposited safely in these ultimate disposal places for many thousands of years, which makes the long-term governance and security of the radioactive waste sites extremely challenging. Nuclear waste also causes widespread social opposition and outrage, as has been observed in Germany since the 1980s, particularly in the Wendland, an area that was proposed for use as a nuclear disposal site. Finally, substantial social and health risks of nuclear energy result from nuclear accidents.

Two key controversies of nuclear energy were exemplified in the nuclear disasters in Chernobyl, Ukraine (part of the former USSR), in 1986 and in Fukushima, Japan, in 2011. Even today, more than 25 years after the nuclear accident in Chernobyl, the affected area – including its water, soil, flora and fauna – is heavily contaminated. There is a debate about the number of human deaths and illnesses, such as thyroid cancer and radiation-induced leukaemia, associated with the nuclear accident. There is also a higher occurrence than usual of children born with birth defects. While the World Nuclear Association (WNA, 2019) argues that only 29 workers have died, the World Health Organisation (WHO, 2005a) argues that about 50 people have died and another 4,000 could eventually die from radiation exposure following the Chernobyl accident.

In 2011, the nuclear power plant at Fukushima was hit by an earthquake scale 9, which was followed by floods caused by a tsunami. This resulted in a nuclear meltdown in several of the reactors. While a few workers died after the accident, the extent of deaths and illnesses caused by the accident is not fully established, although estimates range between 100 and 1,000 expected cancer deaths alone (Caracappa, 2011; Von Hippel, 2011). Following Fukushima, recommendations were made for the nuclear energy sector to better take into account disaster risks, for example, by improving siting procedures and increasing the robustness and design of nuclear reactors (Urban & Mitchell, 2011).

Exercises

1 What health implications are there from energy use, in terms of (a) traditional biomass, (b) fossil fuels and (c) low carbon energy? Try making a list and outlining a few key points for each of these energy sources.
2 Imagine you were an expert advising the World Health Organization on how to reduce adverse health impacts and premature deaths that are linked to the impacts of energy use. What would you advise them? Where would your starting point be to invest in measures that reduce adverse health impacts? What further measures would be helpful? Who would be your target group? You may wish to prepare a short speech or a short PowerPoint presentation.

References

Adams, W. (2000) *The Social Impact of Large Dams: Equity and Distribution Issues.* Thematic Review I.1 prepared as an input to the World Commission on Dams, Cape Town. Available from: https://ideas.repec.org/p/ess/wpaper/id513.html

Asianics Agro-Dev. International (Pvt) Ltd. (2000) *Tarbela Dam and Related Aspects of the Indus River Basin, Pakistan.* A World Commission on Dams (WCD) case study prepared as an input to the WCD, Cape Town. Final Report. Available from: http://s3.amazonaws.com/zanran_storage/www.dams.org/ContentPages/1311315.pdf

Cai, H. & Xie, S. (2011) Traffic-related air pollution modeling during the 2008 Beijing Olympic Games: The effects of an odd-even day traffic restriction scheme. *Science of the Total Environment*, 409 (10), 1935–1948.

Caracappa, F. (2011) *Fukushima Accident: Radioactive Releases and Potential Dose Consequence. Special Session*: The Accident at Fukushima Daiichi – Preliminary Investigations, ANS Annual Meeting, 28 June 2011. Available from: www.ans.org/misc/ FukushimaSpecial-Session-Caracappa.pdf

Gold Standard (2016) *Methodology to Estimate and Verify Averted Disability Adjusted Life Years (ADALYs) from Cleaner Cooking and Cleaner Household Air.* Gold Standard. Available from: www.goldstandard.org/our-work/innovations-consultations/methodologyestimate-and-verify-averted-disability adjusted-life

Haines, A. (2012) Health benefits of a low carbon economy. *Public Health*, 126 (1), 33–39.

Haines, A., McMichael, A.J., Smith, K.R., Roberts, I., Woodcock, J., Markandya, A. & Wilkinson, P. (2010) Public health benefits of strategies to reduce greenhouse-gas emissions: Overview and implications for policy makers. *The Lancet*, 374 (9707), 2104–2114.

Hao, H., Wang, H.W. & Ouyang, M.G. (2011) Comparison of policies on vehicle ownership and use between Beijing and Shanghai and their impacts on fuel consumption by passenger vehicles. *Energy Policy*, 39 (2), 1016–1021.

Hwang, S.S., Xi, J., Cao, Y., Feng, X. & Qiao, X. (2007) Anticipation of migration and psychological stress and the Three Gorges Dam project, China. *Social Science & Medicine*, 65 (5), 1012–1024.

IEA. (2019) *Energy Access*. Paris, International Energy Agency (IEA), OECD/IEA. Available from: www.iea.org/energyaccess/

International Rivers (2008) *The Three Gorges Dam: The Cost of Power*. Available from: www.internationalrivers.org/resources/three-gorges-dam-the-cost-of-power-2651

IPCC. (2007) *Climate Change 2007: Mitigation*. Contribution of Working Group III to the Fourth Assessment Report of the Intergovernmental Panel on Climate Change. [B. Metz, O.R. Davidson, P.R. Bosch, R. Dave, L.A. Meyer (Eds.)] Cambridge University Press, Cambridge, United Kingdom and New York, NY, USA. Available from: www.ipcc.ch/ipccreports /ar4-wg3.htm

IPCC. (2013) Summary for policymakers. In: *Climate Change 2013. The Physical Science Basis*. Contribution of Working Group I to the Fifth Assessment Report of the Intergovernmental Panel on Climate Change (IPCC). [Stocker, T.F., D. Qin, G.-K. Plattner, M. Tignor, S.K. Allen, J. Boschung, A. Nauels, Y. Xia, V. Bex & P.M. Midgley (Eds.)]. Cambridge University Press, Cambridge, United Kingdom and New York, NY, USA. Available from: www.ipcc.ch/ pdf/assessment-report/ar5/wg1/WG1AR5_SPM_FINAL.pdf

La Rovere, E.L. & Mendes, F.E. (2000) *Tucuruí Hydropower Complex, Brazil. A WCD case study* prepared as an input to the World Commission on Dams (WCD), Cape Town. Final Report. Available from: http://citeseerx.ist.psu.edu/viewdoc/download?doi=10.1.1.132.9049&rep=rep1&type=pdf

Majid Cooke, F., Nordensvard, J., Bin Saat, G., Urban, F., Siciliano, G. (2017) The limits of social protection: The case of hydropower dams and indigenous peoples' land. *Asia & the Pacific Policy Studies*. doi: 10.1002/app5.187.

Markandya, A., Armstrong, B.G., Hales, S., Chiabai, A., Criqui, P., Mima, S., Tonne, C. & Wilkinson, P. (2009) Public health benefits of strategies to reduce greenhouse-gas emissions: Low-carbon electricity generation. *The Lancet*, 374(9706), 2006–2015. https://doi.org/10.1016/S0140-6736(09)61715-3.

Morgenstern, V., Zutavern, A., Cyrys, J., Brockow, I., Koletzko, S., Krämer, U., Behrendt, H., Herbarth, O., Von Berg, A., Bauer, C.P., Wichmann, H.E. & Heinrich, J. (2008) Atopic diseases, allergic sensitization, and exposure to traffic-related air pollution in children. *American Journal of Respiratory and Critical Care Medicine*, 177 (12). https://doi.org/10.1164/rccm.200701-036OC

NGO Forum Cambodia (2013) *The Kamchay Hydropower Dam: An Assessment of the Dam's Impacts on Local Communities and the Environment*. NGO Forum, Phnom Penh. North, S. (2012) The Transformation of China's Dirtiest City. City Brand, placesbrands. Available from: http://placesbrands.com/the-transformation-of-filthy-linfen-once-chinasdirtiest-city/

Rao, S., Pachauri, S., Dentener, F., Kinney, P., Klimont, Z., Riahi, K. & Schoepp, W. (2013) Better air for better health: Forging synergies in policies for energy access, climate change and air pollution. *Global Environmental Change*, 2 (5), 1122–1130.

Rissler, P. (2002) *The Three Georges Project on the Yangtze: Reservations and Reality*. DTK German Dam Committee. Available from: http://talsperrenkomitee.de/das_three_gorges_project_am_yangtze/das_three_gorges_project.htm

Seinfeld, J.H. & Pandis, S.N. (2006) *Atmospheric Chemistry and Physics: From Air Pollution to Climate Change*. 2nd edition. Hoboken, NJ, Wiley.

Siciliano, G., Urban, F., Kim, S. & Lonn, D.P. (2015) Hydropower, social priorities and the rural-urban development divide: The case of large dams in Cambodia. *Energy Policy*, 86 (11), 273–285.

Smith, K.R., Woodward, A., Campbell-Lendrum, D., Chadee, D.D., Honda, Y., Liu, Q., Olwoch, J.M., Revich, B. & R. Sauerborn (2014) Human health: impacts, adaptation, and co-benefits. In: *Climate Change 2014: Impacts, Adaptation, and Vulnerability*. Part A: Global and Sectoral Aspects. Contribution of Working Group II to the Fifth Assessment

Report of the Intergovernmental Panel on Climate Change [Field, C.B.,V.R. Barros, D.J. Dokken, K.J. Mach, M.D. Mastrandrea, T.E. Bilir, M. Chatterjee, K.L. Ebi, Y.O. Estrada, R.C. Genova, B. Girma, E.S. Kissel, A.N. Levy, S. MacCracken, P.R. Mastrandrea, and L.L. White (Eds.)]. Cambridge University Press, Cambridge, United Kingdom and New York, NY, USA, pp. 709–754. Available from: www.ipcc.ch/report/ar5/wg2/

Smithson, P., Addison, K. & Atkinson, K. (2008) *Fundamentals of the Physical Environment.* 4th edition. London, Routledge.

Urban, F. & Mitchell, T. (2011) *Climate Change, Disasters and Electricity Generation.* Institute of Development Studies (IDS), Brighton. Strengthening Climate Resilience Discussion Paper 8. Available from: www.ids.ac.uk/files/dmfile/UrbanAndMitchell_2011_Electricity DP82.pdf

Urban, F., Nordensvärd, J., Khatri, D. & Wang, Y. (2013) An analysis of China's investment in the hydropower sector in the Greater Mekong Sub-Region. *Environment, Development and Sustainability,* 15 (2), 301–324.

US NRC (2002) *Radioactive Waste: Production, Storage, Disposal.* US Nuclear Regulatory Commission. NUREG/BR-0216, Rev. 2. Available from: www.nrc.gov/docs/ ML1512/ ML15127A029.pdf

Von Hippel, F.N. (2011) The radiological and psychological consequences of the Fukushima Daiichi incident. *Bulletin of the Atomic Scientists,* 67 (5), 27–36.

Walsh, B. (2007) The World's Most Polluted Places. Linfen, China. Time. Available from: http://content.time.com/time/specials/2007/article/0,28804,1661031_1661028_ 1661016,00.html

WEC (2018) *Energy Resources.* Available from: www.worldenergy.org/data/ resources/

WHO. (2000) *Addressing the Links between Indoor Air Pollution, Household Energy and Human Health.* Geneva, World Health Organization (WHO).

WHO. (2005a) *Chernobyl: The True Scale of the Accident.* Geneva, World Health Organization (WHO). Joint News Release WHO/IAEA/UNDP. Available from: www.who.int/ mediacentre/news/releases/2005/pr38/en/index.html

WHO. (2005b) *Household Air Pollution and Health.* Updated February 2016. Geneva, World Health Organization (WHO). Factsheet No 292. Available from: www.who.int/ mediacentre/factsheets/fs292/en/

WHO. (2006) *Fuel for Life: Household Energy and Health.* Indoor air pollution. Geneva, World Health Organization (WHO). Available from: www.who.int/indoorair/publications/ fuelforlife/en/index.html

WHO. (2016) *Health Statistics and Information Systems.* Geneva, World Health Organization (WHO). Available from: www.who.int/healthinfo/global_burden_ disease/metrics_daly/ en/

WHO. (2019) *Air Pollution.* Available from: https://www.who.int/airpollution/en/

WNA. (2019) *World Nuclear Power Reactors & Uranium Requirements.* World Nuclear Association (WNA). Available from: www.world-nuclear.org/information-library/ facts-and-figures/world-nuclear-power-reactors-archive/reactor-archive-march-2012.aspx

Yang, S. (2011) *Wood Smoke from Cooking Fires Linked to Pneumonia, Cognitive Impacts.* News Center, UC Berkeley. Available from: http://newscenter.berkeley.edu/ 2011/11/10/ cookstove-smoke-pneumonia-iq/

Yu, X. (2003) Regional cooperation and energy development in the Greater Mekong Sub-Region. *Energy Policy,* 31 (12), 1221–1234.

9

THE SOCIAL IMPLICATIONS OF ENERGY AND DEVELOPMENT

Social opportunities: improving living standards and well-being by energy access

Overcoming fuelwood collection

As we discussed earlier in this book, fuelwood collection is a time-intensive and labour-intensive burden that is predominantly undertaken by women and children, mainly girls, in developing countries. At the same time, fuelwood collection is often a safety risk. Women and children face several severe safety risks when leaving their homes in search of fuelwood; for example, violence towards them (including sexual violence), theft and even murder (Urban & Lind, 2011). More details on this topic are explored in the section of this chapter titled 'Gender perspectives'. Reducing the need for fuelwood collecting by providing access to modern energy, such as electricity, natural gas, biogas and solar energy for cooking, heating and lighting, can therefore have a direct positive effect on the living standards and well-being of affected individuals.

Lighting and education

One of the main social opportunities of providing modern energy access is to generate light for people who would otherwise be left in the dark. It is reiterated by researchers, governments, non-governmental organisations (NGOs) and UN institutions alike that this contributes to improved education. The logic of this reasoning is that access to modern energy can generate lighting for children (and adults) to study after dark in their homes and in community buildings and that it can provide electricity to schools and other education institutions for lighting and communication (e.g. radios, laptops/computers, TVs, mobile phones, tablets).

In recent years, an increasing number of initiatives, projects and programmes have emerged for bringing solar lighting to the energy poor, particularly in sub-Saharan Africa and in India.

Gustavsson (2007) examined the educational benefits of solar electric services in Zambia. The research finds that there are benefits for schoolchildren in terms of providing lighting for education; however, there are no insights on the actual improvements in schoolchildren's marks, as they were not measured. This is a common problem: The benefits of modern energy for lighting and its educational benefits are often reiterated, and while the hours of studying after the dark have been reported to go up, there is very little data available to link educational achievements with access to lighting.

Income-generating activities and productive use of energy

It is often suggested by researchers, donors, policymakers and practitioners that there is a link between access to modern energy and income-generating activities. Energy is a prerequisite for powering economic activities such as agriculture, industries, services, construction, etc. Energy is also needed to power transport, telecommunications and medical services (Goldemberg & Lucon, 2009; Urban, 2014).

Access to reliable and affordable energy services means that local entrepreneurs and companies can generate local employment and income and contribute to local development, in rural as well as urban areas (GIZ & EUEIPDF, 2011). There is a wide range of income-generating activities at the local level that need access to modern energy, such as for agricultural production and food processing, handicrafts, shop-keeping, communications and entertainment. Electricity and natural gas can contribute to these income-generating activities, as can renewable energy such as solar, wind, hydropower and biogas. The following example explains how modern energy access can contribute to income-generating activities.

In the Kalalé district of rural Benin, a low-income country in West Africa, farmers' agricultural output depends on adequate irrigation. Climate change has led to water stress and drought in Benin, which makes agriculture difficult, particularly in the dry season. Thanks to a project by Stanford University and the Solar Electric Light Fund, villages use solar-powered pumps for irrigation (SELF, 2007). The photovoltaic (PV) systems power a water pump and a drip-irrigation system. Farmers can now grow vegetables for their own use and sell them on the local markets to generate a higher income, even in the dry season (Woods Institute, 2010). In addition, the PV systems provide electricity to local community buildings, such as schools and clinics (Urban & Nordensvärd, 2013).

Drawing on quantitative data on a national level, Yoo and Kwak (2010) found a relationship between electricity consumption and economic growth in seven Latin American countries. Their study found that electricity consumption directly affects economic growth in Argentina, Brazil, Chile, Columbia, Ecuador and Venezuela. No clear trend could be found for Peru. The data show that access to electricity can promote economic growth and income generation at an aggregate national level.

Obermaier et al. (2012), on the other hand, examined the link between electricity access and income generation at the local level in poor rural communities in Brazil. While the research found an increase in electricity consumption, an increase in rural incomes could not be observed in the short term. This is an interesting finding compared to the aggregate national-level data of Yoo and Kwak (2010) suggesting a link between access to electricity and incomes in Brazil. Similar conclusions were drawn by Kooijman-van Dijk (2012), who analysed the link between access to modern energy and income generation at a local level in the Indian Himalayas. The study finds that the uptake of electricity is high, but the link to income generation is low for small informal enterprises. The paper highlights the role of energy access for well-being and improving the living conditions of the rural population. Recent research by Byrne et al. (2014) suggests that small-scale, bottom-up renewable energy investments aimed at local communities are far better at reaching the poor, contributing to poverty alleviation and improving the living standards of the poor than large-scale, top-down renewable energy investments such as the Clean Development Mechanism (CDM).

Productive use of energy refers narrowly to (a) using energy for income-generating purposes and more broadly to (b) using energy for a wider set of welfare-related activities such as healthcare and education. It is often mentioned in relation to electricity as well as renewable energy (Kapadia, 2004). We discussed income generation earlier; this section therefore focusses on a wider set of welfare-related activities, such as healthcare. Solar-powered refrigerators, clinics and healthcare centres are nowadays found almost everywhere in rural and remote locations that lack access to the central grid in the developing world. Sometimes these health services are powered by hybrid systems, such as solar PV and wind, hydro or biomass, to ensure maximum service. One of the most innovative ways of providing healthcare and medical services are solar-powered refrigerators for storing vaccinations that are transported by camels to rural communities in Kenya and Ethiopia. This solar-powered medical system was produced by Princeton Engineering.

Critique using quantitative evidence

There is a consensus among researchers, development practitioners and policymakers that access to modern energy improves the lives and livelihoods of people living in energy poverty. While this is a commonly accepted finding and certainly holds true with regard to well-being and living conditions, quantitative evidence for specific indicators (such as impacts on economic growth or income generation) is more complex. Some of these assumptions are contentious and disputed. We will therefore look at some data and evidence to support and/or critique some of these assumptions and findings. We discussed earlier that there is a link between income levels and energy access. A correlation has been found between rising income levels, both at household level and at national level, and rising energy access. Consequently, countries that have higher incomes tend to have higher electricity access rates (Urban & Nordensvärd, 2013). While a clear trend can be seen, there are always exceptions. An exception is, for example, Vietnam, which is a lower middle-income country, but has an electrification rate of about 100%. This is due to consolidated government efforts over many years for rural and urban electrification. Quantitative data from Asia shows that countries that have higher per capita incomes, such as Malaysia, China and Thailand, have higher electrification rates, whereas countries with lower per capita incomes, such as Cambodia and Bangladesh, have lower electrification rates. The data indicate there may be a cut-off point at approximately US$15,000–20,000 GDP/capita PPP (purchasing power parity) when electrification rates are about 100%. Vietnam is an obvious outlier, with an average GDP/capita PPP around US$7000, yet an electrification rate of nearly 100% in 2018 (IEA, 2019; World Bank, 2019). This is due to concerted government efforts for rural and urban electrification, as well as a rapid expansion of domestic hydropower plants. A similar strategy for electrification was used in China several decades ago. Sri Lanka, Bhutan, Nepal and Afghanistan are other countries that still have relatively low per capita incomes, however electrification rates have rapidly increased in recent years to around 95%. The SE4All initiative can therefore show significant improvements for electrification rates across developing Asia. The energy access situation for Latin America, North Africa and the Middle East is comparable to that in developing Asia. Sub-Saharan Africa is another case. A large number of countries are very poor, well below US$3000 or 4000 GDP/capita PPP. Those that have higher incomes still have lower electrification rates than similar countries in other developing regions around the world, such as Nigeria (about 55% electrification rate in 2017), Namibia (about 52% electrification rate in 2017) or Angola (about 40% electrification rate in 2017). Even South Africa has an electrification rate of only about 85% (IEA, 2019; World Bank, 2019). The different situation in sub-Saharan Africa might partly be explained by recent, as well as historic, violent conflicts and war. Another reason might be the high prevalence of fragile states, where governments are weak and struggle to adequately govern the country, as well as a lack of sufficient investment in energy infrastructure and continued social inequality (e.g. in South Africa). In all regions of the developing world, rural energy access tends to be lower than urban energy access.

Access to modern energy, such as electricity and clean cooking fuels, also tends to correlate with the Human Development Index (HDI) and the Energy Development Index (EDI). The EDI is a measure calculated by the International Energy Agency (IEA); however, recently it has not been used, and the latest data available is from 2012.

The United Nations' HDI is an indicator that shows how developed a country is. It takes into account the Gross National Income (GNI) per capita, life expectancy and an education index that is composed of average education level and expected length of schooling in the country. OECD countries, particularly the Nordic countries such as Norway, have traditionally been very high in the HDI ranking. Sub-Saharan countries, particularly those affected by armed conflict, have traditionally been low in the HDI ranking. The IEA's EDI is an indicator that shows how developed a country is in energy terms. The EDI takes into account per capita commercial energy consumption (excluding traditional biofuels that were not purchased), per capita electricity consumption in the residential sector, the share of modern fuels in the total residential sector energy use and the share of the population with access to electricity. For each of these four indicators, a separate index is used according to the formula below:

Indicator = (actual value − minimum value) / (maximum value − minimum value)

Performance in each indicator ranges between 0 and 1, with 1 being the highest value. The EDI is then calculated as the mean of these four indicators. The EDI is only calculated for developing countries; as for developed countries, the EDI would be 1. The highest-ranking countries are typically located in the Middle East and Latin America. The lowest-ranking countries are typically located in sub-Saharan Africa. Based on 2012 data, the country with the lowest EDI of 0.01 is the Democratic Republic of Congo (DRC), which has an electrification rate of 15% (World Bank, 2019). Correlating with the EDI is the HDI, which also ranks the DRC at the bottom (in the same position as Niger) with an HDI of 0.304 (UNDP, 2014). Honduras ranks in the statistical middle of the EDI ranking, with an EDI of 0.285 and ranks number 120 out of 192 countries for the HDI, a little below the statistical middle. Libya and Iran are at the top of the EDI ranking with scores of 0.923 and 0.880, respectively. Yet, Libya ranks 67th and Iran ranks 76th in the HDI. The top HDI places are taken by OECD countries, most importantly Norway (ranked first for several years in a row), and these countries also have the highest level of EDI measured at 1 (or 100%). Since 2012, outbreaks of violent conflict and war have influenced the performance of some of these countries, resulting in lower HDIs, such as for Yemen which is in last position in 2019 (UNDP, 2019). In recent years, more sophisticated indicators have been developed, such as the Global Multidimensional Poverty Index (MPI) and the Sustainable Energy Development Index (SEDI) (Iddrisu & Bhattacharyya, 2015).

We can clearly see, however, that there is a correlation between energy development, measured in form of the EDI, and human development, measured in the form of the HDI. Those countries that have very low human development scores have also very low energy development scores; those countries that have average human development scores have average energy development scores; and those countries that have high human development scores have also high energy development scores.

The findings mentioned above are supported by Yoo and Kwak (2010), who find a relationship between national electricity consumption and national economic growth in seven Latin American countries. They argue that electricity consumption directly affects economic growth in Argentina, Brazil, Chile, Columbia, Ecuador and Venezuela. At the local level, these findings could not be confirmed for Brazil (see Obermaier et al., 2012). Nevertheless, these findings highlight the role of energy access for well-being and improving the living conditions of the rural population.

Gender perspectives

Gender, energy and development

Gender plays an important role for energy and development. It is women and girls who are mainly responsible for organising access to energy, such as collecting fuelwood, using energy such as for cooking and making energy choices such as what fuel to use, how much, where to get it from and at what price. As Table 9.1 shows, women and girls in rural households are usually responsible for decisions about which fuel to use, which cooking device to use, what size of cooking device to use and how many, what type and quantity of food to cook. Women and girls therefore usually make most of the decisions about household energy.

At the same time, women and girls are disproportionately affected by a lack of modern energy access and the social implications this has, such as with regards to health. The FAO (2006: 1) particularly highlights the role of gender for the world's poorest women living in rural areas in developing countries. They depend

TABLE 9.1 Responsibility of women and men for household energy choices

Component	Decision about . . .	Decision typically made by . . .	Variables influencing the decision-making
Kitchen	• Location of kitchen • Construction material of kitchen • Layout of kitchen	• Men • Men • Women	• Economic status • Climate • Availability of construction material • Secondary use of kitchen
Fuel	• Which fuel to use	• Men/women	• Availability in terms of costs and distance • Combustion characteristics • Taste • Convenience
Device (e.g. stove)	• Which device to use • What size • How many	• Men • Women • Women	• Economic status • Task suitability • fuel type • Awareness level • Cooking cycle • Time availability • Culture, inhibitions etc.
Vessel (e.g. cooking pot)	• Material • Size • Shape	• Women • Women • Women	• Economic status • Tradition • Food habits
Food	• Type • Quantity	• Women • Women	• Habits • Economic status

Source: Adapted from FAO (2006: 27)

on subsistence farming for providing food for their families, and 'are disproportionately affected by the lack of modern fuels and power sources for farming, household maintenance and productive enterprises'. The FAO (2006) further reports that:

> Poor women in rural areas of developing countries generally have a more difficult time compared to men, due to their traditional socio-cultural roles. They often spend long hours collecting fuel wood and carrying it back home over long distances. The time and labour expended in this way exhausts them and limits their ability to engage in other productive and income-generating activities. Their health suffers from hauling heavy loads of fuel and water, and from cooking over smoky fires. Their opportunities for education and income generation are limited by lack of modern energy services, and as a result their families and communities are likely to remain trapped in poverty.
>
> *FAO (2006: 2)*

Fuelwood collection alone can take several hours each day and involve distances of several kilometres. A study from the Sahel found that women travelled an average 15–20 km per day to collect fuelwood. Fuelwood is scarce in this region due to deforestation and desertification, causing even more pressure on women's lives (FAO, 2006). However, the problem is not only confined to women: Children, especially girls, are often encouraged to stay at home rather than to attend school to help their mothers with fuelwood collection, cooking and other household and care tasks. The strains from fuelwood collection and the adverse health effects from indoor air pollution tend to come in addition to a serious of disadvantages for women and girls, such as poorer health, lower nutritional input and poorer diets, lower education and high levels of illiteracy, low or no incomes and financial opportunities, exhausting fuel wood and water collection duties, in addition to household, childcare and family duties, safety risks associated with fuelwood and water collection, lower social status (sometimes even social stigma) and lack of power, etc. The FAO therefore recommends a series of gender concerns to be integrated in energy and rural development plans, for example, gender-mainstreaming at the organisational, programme and policy level, use of analytical tools such as gender analysis when planning policies and programmes and inclusion of gender indicators in impact assessments (FAO, 2006). In line with these suggestions, the World Bank supports energy pilot projects in Senegal, Mali, Benin, Tanzania and Kenya that are mainstreaming gender. The World Bank emphasises that improving energy equality and social inclusion in energy projects improves overall development outcomes (ESMAP, 2014).

There are other innovative approaches for increasing the emancipation of women and girls with regard to energy. Let's take the case study of the solar cooker project in Chad as an example. The project is based in a refugee camp for those who have fled from the civil war in Darfur, Sudan, to neighbouring Chad. It is estimated that about 2.5 million people have been displaced by the war. In these refugee camps, it is the responsibility of women and girls to collect fuelwood. In doing so they 'have to leave the safety of their camps and risk rape, assault, abduction, theft, exploitation or even murder' (Urban & Lind, 2011: 28). The Women's Commission for Refugee Women and Children (WCRWC) recommends that more renewable energy, such as solar energy, is being used to replace fuelwood collection by women and girls. The Solar Cooker Project of the United Nations High Commissioner for Refugees (UNHCR), the Dutch foundation KoZon and Cooperative for Assistance and Relief Everywhere (CARE) provide solar cookers to female refugees. In 2006, the project started supplying solar cookers to 7,000 people in the refugee camps Iridimi and Oure Cassoni in Chad. In 2011, the initiative was led by Jewish World Watch (JWW) in refugee camps Iridimi, Touloum and Oure Cassoni and benefited about 60,000 people, according to Urban & Lind (2011).

The women and girls use the solar cookers thereby avoiding the need for fuelwood collection. They are also being trained to make the solar cookers themselves and are paid a small amount of money for doing so. This is a welcome option for women and girls that are often illiterate and have lost land and access to their farming incomes due to displacement. 'Each family in the refugee camp gets two solar cookers distributed by the NGO for free. Every women and girl of the age of 15 or older gets trained how to use and produce a solar cooker' (Urban & Lind, 2011: 29). There is a local manufacturing facility for the solar cookers, which are made of cardboard and foil. Local production of the cookers makes them sustainable and viable in the long run.

> A survey from Iridimi refugee camp showed that more than 70% of the surveyed women and girls collected fuel wood between 3 [and] 7 times per week. Since using the solar cooker more than half of the women and girls do not collect fuel wood at all, while about a quarter collect fuel wood only once a week.
>
> *(Urban & Lind, 2011: 29)*

There are a few practical and cultural disadvantages, such as the short lifespan of the solar cooker which needs renewing every few months, longer cooking times compared to cooking over open fire and prevailing cultural cooking practices which have to be changed. However, these disadvantages are marginal in relation to the advantages of the solar cookers: keeping women and girls safe, avoiding the time-consuming drudgery of fuelwood collection, avoiding the adverse health impacts from cooking with fuelwood, women and girls earning an income and being trained technically, reducing pressure on forests and woodlands and using low carbon energy (Urban & Lind, 2011).

Exercises

1   Imagine you were an expert on gender and development in the energy field. Write down a few recommendations for policymakers: What role does gender play for energy and development? How can women and girls be

empowered with regards to energy supply and energy use? What are the practical implications?

2   Imagine you were an advisor to the government in a developing country. First, find out how many people in your country of choice do still not have access to electricity and clean cooking options. Check the IEA energy access database: www.iea.org/energyaccess/database/. Second, what would you suggest to the government to increase the living standards and the well-being of the local population in relation to energy access? What role does energy for lighting, productive uses and other functions play? Come up with a list of recommendations to the government.

References

Byrne, R., Ockwell, D., Urama, K., Ozor, N., Kirumba, E., Ely, A., Becker, S. & Gollwitzer, L. (2014) *Sustainable Energy for Whom? Governing Pro-poor, Low-carbon Pathways to Development: Lessons from Solar PV in Kenya*. STEPS Centre, Brighton Steps Working Paper 61. Available from: http://steps-centre.org/wp-content/uploads/Energy-Access-online.pdf

ESMAP. (2014) *Gender: Social Inclusion in the Energy Sector*. Energy Sector Management Assistance Programme (ESMAP). The World Bank. Available from: www.esmap.org/EnergyandGender

FAO. (2006) *Energy and Gender in Rural Sustainable Development*. Lambrou, Y. & Piana, G. (Eds.). Rome, Food and Agriculture Organization of the United Nations (FAO). Available from: www.fao.org/docrep/010/ai021e/ai021e00.htm

GIZ. & EUEIPDF. (2011) *Productive Use of Energy (PRODUSE)*. Duetsche Gesellschaft fur Internaionale Zusammenarbeit (GIZ) and EU Energy Initiative Partnership Dialogue Facility (EUEIPDF). Available from: www.euei-pdf.org/thematic-studies/productive-use-ofenergy-produse

Goldemberg, J. & Lucon, O. (2009) *Energy, Environment and Development*. 2nd edition. Oxon, Earthscan, Routledge.

Gustavsson, M. (2007) Educational benefits from solar technology – access to solar electric services and changes in children's study routines, experiences from eastern province Zambia. *Energy Policy*, 35 (2007), 1292–1299.

Iddrisu, S. & Bhattacharyya, S.C. (2015) Sustainable Energy Development Index: A multi-dimensional indicator for measuring sustainable energy development. *Renewable and Sustainable Energy Reviews,* 50(10), 513–530.

IEA. (2019) *Statistics*. Available from: https://www.iea.org/statistics/

Kapadia, K. (2004) *Productive Uses of Renewable Energy*. A Review of Four Bank-GEF Projects. Washington DC, The World Bank. Available from: http://ww.martinot.info/Kapadia_WB.pdf

Kooijman-van Dijk, A.L. (2012) The role of energy in creating opportunities for income generation in the Indian Himalayas. *Energy Policy*, 41, 529–536.

Obermaier, M., Szklo, A., La Rovere, E.L. & Pinguelli, R. (2012) An assessment of electricity and income distributional trends following rural electrification in poor northeast Brazil. *Energy Policy*, 49, 531–540.

SELF. (2007) *Solar Drip Irrigation Project – Benin, West Africa*. [Video]. Solarnow2007. Solar Electric Light Fund (SELF). Duration 4:08 minutes. Available from: www.youtube.com/watch?v=RTtBEbf-NRs

UNDP. (2014) Human Development Report 2014: Sustaining Human Progress. Reducing Vulnerabilities and Building Resilience.

UNDP. (2019) United Nations Development Programme, Global Human Development Indicators. http://hdr.undp.org/en/countries.

Urban, F. (2014) *Low Carbon Transitions for Developing Countries*. Oxon, Earthscan, Routledge.

Urban, F. & Lind, J. (2011) Low carbon energy and conflict: A new agenda. *Boiling Point*, 59 (2011), 28–29.

Urban, F. & Nordensvärd, J. (2013) *Low Carbon Development: Key Issues*. Oxon, Earthscan, Routledge.

Woods Institute (2010) *Stanford Study: Solar Irrigation in Africa*. [Video]. Stanford University. Duration 4:54 minutes. Available from: www.youtube.com/watch?v=56VzLXnIqEQ

World Bank. (2019) *Open Data*. Available from: https://data.worldbank.org/

Yoo, S-H. & Kwak, S-Y. (2010) Electricity consumption and economic growth in seven South American countries. *Energy Policy*, 38 (1), 181–188.

10

ENVIRONMENTAL IMPLICATIONS – ENERGY USE AND CLIMATE CHANGE

Climate change

Energy and the basics of climate science

This part builds on issues we discussed in Chapter 1, providing more detail.

Defining climate change

There are two commonly used definitions of climate change, one more narrowly defined by the United Nations Framework Convention on Climate Change (UNFCCC) and one more broadly defined by the Intergovernmental Panel on Climate Change (IPCC). The UNFCCC defines climate change as 'a change of climate which is attributed directly or indirectly to human activity that alters the composition of the global atmosphere and which is in addition to natural climate variability observed over comparable time periods' (UNFCCC, 1992: 3). The UNFCCC therefore makes a distinction between 'climate change', which is regarded as a purely anthropogenic phenomenon caused by human activities altering the composition of the atmosphere, and 'climate variability', which is attributed to natural causes (UNFCCC, 1992; IPCC, 2007). On the other hand, the IPCC (2007: 30) defines climate change as

> a change in the state of the climate that can be identified (eg using statistical tests) by changes in the mean and/or the variability of its properties, and that persists for an extended period, typically decades or longer. It refers to any change in climate over time, whether due to natural variability or as a result of human activity.

In this definition, the IPCC acknowledges that climate change has both natural and human causes; thus, climate change may result from natural processes, such as variations in the earth's orbit, as well as from anthropogenic changes, such as alterations in the composition of the atmosphere or of land-use patterns (IPCC, 2007, 2014a).

The link between energy and climate change

As we discussed in Chapter 1, about 80% of the global primary energy supply comes from fossil fuels, mainly oil and coal (IEA, 2019). Fossil energy resources are limited, and fossil energy use is associated with a number of negative environmental effects, most importantly global climate change (Quadrelli & Peterson, 2007) but also natural resource depletion and air pollution. The next section discusses how energy use and climate change are linked. The IPCC estimates that about 70% of all greenhouse gas (GHG) emissions worldwide come from energy-related activities. This is mainly from fossil fuel combustion for heat supply, electricity generation and transport and includes carbon dioxide (CO_2), methane and some traces of nitrous oxide (IPCC, 2007). It is well documented that these emissions contribute to global climate change. Energy use has potentially significant climate impacts, which are assumed to exceed the impacts from other sources, such as land use and other industrial activities. It is therefore considered crucial to transition from fossil fuels to renewable energy, as well as to increase energy efficiency. Historically, climate change has mainly been caused by today's industrialised countries.

The IPCC states that the

> atmospheric concentrations of carbon dioxide, methane, and nitrous oxide have increased to levels unprecedented in at least the last 800 000 years. Carbon dioxide concentrations have increased by 40% since pre-industrial times, primarily from fossil fuel emissions and secondarily from net land use change emissions.
>
> *(IPCC, 2014a: 7)*

The concentrations of GHGs have increased rapidly, with concentrations of carbon dioxide, methane, and nitrous oxide exceeding the pre-industrial levels by about 40%, 150% and 20%, respectively, in 2011 (IPCC, 2014a). Energy use from fossil fuels therefore contributes directly to GHG emissions and thereby contributes to rising temperatures and other observed effects of global climate change. The IPCC has an extremely high confidence level of 95% probability that global climate change is anthropogenic, caused by excessive GHG emissions (IPCC, 2014a, 2014b). At the global scale, the atmospheric concentration of carbon dioxide has increased from a pre-industrial value of approximately 280 parts per million (ppm) to around 410 ppm in 2019 (NASA, 2019). Go back to Figure 3.1 for an overview of rapidly increasing global CO_2 emissions in the atmosphere. Globally, annual emissions of carbon dioxide have risen dramatically in recent decades, reaching around 25 Gt (gigatonnes) of carbon dioxide in 2005, around 30 Gt in 2010 and around 40 Gt in 2018 (IPCC, 2007, 2014a, 2018).

The greenhouse effect

So far, we have discussed the link between energy use and GHG emissions, particularly carbon dioxide, but we have not discussed the mechanisms that link GHGs and climate change. This has to do with the so-called greenhouse effect. We will first discuss the natural greenhouse effect and afterwards we will discuss the anthropogenic (or enhanced or human-induced) greenhouse effect.

The natural greenhouse effect

The atmosphere plays a critical role in the heat budget of the earth. The outgoing terrestrial radiation of heat is of a longer wavelength than the incoming solar radiation. This difference is vitally important for the climate because many of the gases of the atmosphere absorb energy selectively. Relatively little of the incoming solar radiation is absorbed by the gases and particles of the atmosphere but, in contrast, a much greater proportion of the outgoing terrestrial radiation – which is of longer wavelength – is absorbed by the atmosphere, with the result that heat is retained in the atmosphere.

The overall absorption of outgoing terrestrial radiation by the atmosphere depends upon its composition, as some gases absorb particular wavelengths more than other gases. This phenomenon of heating of and by the atmosphere is known as the greenhouse effect. The greenhouse effect is an important part of the earth's heat budget, and thus the heat budget is affected by the composition of the atmosphere (see Figure 10.1).

The greenhouse effect is called this way because the heat is trapped close to the Earth's surface. Like a greenhouse keeps heat under its glass panels. The layer of greenhouse gases acts like a blanket that covers the Earth to retain heat.

The gases that absorb some of the energy radiated by the Earth's surface are called greenhouse gases.

The natural greenhouse effect occurs directly as a result of the composition of the atmosphere and the radiative properties of its molecules, such as water vapour. The natural greenhouse effect is not a result of human activities. Rather it occurs because of the Earth's atmosphere that contains gases that absorb energy. Even in the absence of any human impact on the atmosphere, the natural greenhouse effect keeps the temperature of the Earth at about 20 °C warmer than would otherwise be the case in the absence of the greenhouse effect (Barry & Chorley, 2003; Houghton, 2009). The natural greenhouse effect has been an important prerequisite in allowing the evolution of life on Earth.

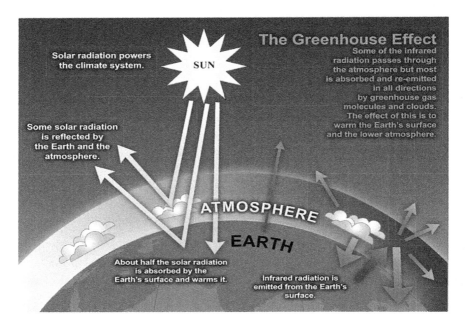

FIGURE 10.1 The earth's heat budget and the greenhouse effect

Source: IPCC, 2014

GHGs and the anthropogenic greenhouse effect

The anthropogenic greenhouse effect (or enhanced or human-induced green-house effect) is caused by the emissions of greenhouse gases (GHGs). The major six anthropogenic GHGs that are covered under the UNFCCC (1998) are as follows:

- carbon dioxide (CO_2) – mainly released from the combustion of fossil fuels as well as from the destruction of carbon sinks (e.g. deforestation and land use change)
- methane (CH_4) – mainly released from agriculture
- nitrous oxide (N_2O) – mainly released from agriculture
- hydrofluorocarbons (HFCs) – mainly released from industrial processes
- perfluorocarbons (PFCs) – mainly released from industrial processes
- sulfur hexafluoride (SF_6) – mainly released from industrial processes

Water vapour is an important GHG too, but it occurs naturally in the atmosphere. Anthropogenic GHGs involve emissions from human sources, such as fossil fuel combustion, deforestation and agricultural practices. CO_2 is the most frequently occurring and therefore most important GHG (e.g. having the highest concentration of GHG in the atmosphere), although it has a relatively low ability to absorb radiation in comparison to other GHG.

The importance of different gases in promoting global warming depends upon:

- the concentration of the GHG in the atmosphere
- the GHG's global warming potential (GWP), that is the amount of radiation they can adsorb
- the length of time that the GHG stay in the atmosphere.

Calculating the global warming potential (GWP) of different gases in comparison with the GWP of carbon dioxide, measured in carbon dioxide equivalents (CO_2-equivalent or CO_2e), allows us to understand their impact comparatively (Houghton, 2009).

As long as the atmospheric concentrations of GHGs remain constant, there is no imbalance in the radiative budget for the Earth, and hence no radiative forcing and no climate change. However, any change in the abundance of GHGs is likely to cause radiative forcing and climate change. The most important cause of anthropogenic climate change is the accumulation of GHGs in the atmosphere, especially the accumulation of carbon dioxide, such as through the combustion of fossil fuels, land use changes, deforestation and industrial processes.

Energy from non-fossil fuels, such as renewable energy and lower carbon energy, is therefore crucial for mitigating the GHG emissions that lead to climate change. Other strategies, such as reducing energy use and increasing energy efficiency, are also required.

The impacts of global climate change

In this section, we will discuss some of the observed and expected impacts of climate change according to the IPCC Fourth and Fifth Assessment reports. While climate change has mainly been caused by today's industrialised countries, it affects the world's poorest the worst. This is due to their high vulnerability and their limited resources to adapt to climatic impacts.

According to the IPCC (2014a, 2014b), the global mean surface temperature has risen by about 1 °C degree (likely range 0.8–1.2 degrees C) since pre-industrial times (IPCC, 2018). This increase has been particularly significant over the last 50 years. The political target is to limit global warming to between 1.5 to 2 degrees C, as suggested by the Paris Agreement. The Paris Agreement in December 2015 set out a global action plan to limit global warming to well below 2 °C – and pursue efforts to limit the temperature increase to 1.5 °C. The IPCC (2018) published a report in late 2018 on how to achieve the 1.5 degree target. The IPCC 1.5 °C highlights how dangerous climate change can be avoided from a sociotechnical, economic and policy perspective. Yet, the IPCC warns that if no action is taken beyond the current level, global atmospheric temperature increases will reach 1.5 °C between 2030 and 2052 rather than in 2100 (IPCC, 2018).

Therefore, the 1.5 °C target is becoming increasingly difficult to achieve. The majority of climate scientists agree that the possibility of staying below the 2 °C rise threshold between 'acceptable' and 'dangerous' climate change by 2100 is becoming less likely as no serious global action on climate change is taken (Richardson et al., 2009; IPCC, 2018). Some countries have made considerable progress in mitigating emissions between 1990 and today, such as Sweden, Denmark, Germany, France, Italy, the UK, Russia and other former Soviet countries. Yet, other countries have hugely increased their emissions during this time, such as the USA, Canada, Australia, China, India, South Africa, Mexico, Brazil, Indonesia, Malaysia, Saudi Arabia, Qatar and the United Arab Emirates, to name just a few. Some of these countries are emerging economies that are still in the process of development, whereas others are industrialised countries such as the USA, Canada and Australia that have not managed to transform their energy systems and other carbon-intensive industries to mitigate their emissions.

Climate scientists estimate that for a 50% chance of achieving the 2 °C target, a global atmospheric carbon dioxide equivalent concentration of 400–450 ppm should not be exceeded (Richardson et al., 2009), which would require an immediate reduction of global GHG emissions to about 60%–80% by 2100 (Richardson et al., 2009). Nevertheless, the 400 ppm target has been exceeded recently (NASA, 2019) and still global emissions are rising. There is therefore a need to reduce emissions rapidly and significantly to avoid dangerous climate change (IPCC, 2018). Given the current state of global emissions, the Paris Agreement implicitly implies that the 1.5 to 2 degree C target will need to be achieved using advanced mitigation technologies such as BECCS (bioenergy carbon capture and storage).

See Figure 10.2 for cumulative carbon dioxide emissions so far and Figure 10.3 for required mitigation pathways to achieve the 1.5 degree global target.

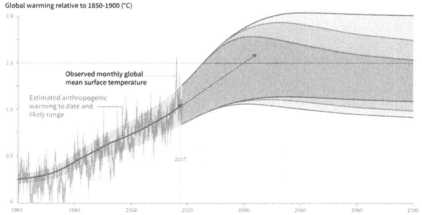

a) Observed global temperature change and modeled responses to stylized anthropogenic emission and forcing pathways

Global warming relative to 1850-1900 (°C)

FIGURE 10.2 Cumulative carbon dioxide emissions and future emission scenarios

Source: IPCC, 2018

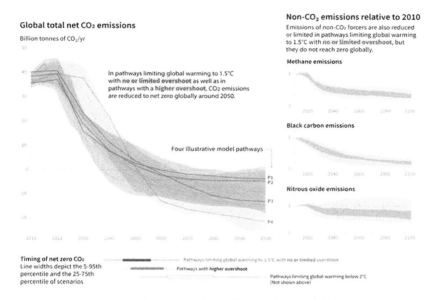

Global total net CO₂ emissions

Billion tonnes of CO₂/yr

In pathways limiting global warming to 1.5°C with no or limited overshoot as well as in pathways with a higher overshoot, CO₂ emissions are reduced to net zero globally around 2050.

Four illustrative model pathways

P1
P2

P3

P4

Timing of net zero CO₂
Line widths depict the 5-95th percentile and the 25-75th percentile of scenarios

Pathways limiting global warming to 1.5°C with no or limited overshoot

Pathways with higher overshoot

Pathways limiting global warming below 2°C (Not shown above)

Non-CO₂ emissions relative to 2010
Emissions of non-CO₂ forcers are also reduced or limited in pathways limiting global warming to 1.5°C with no or limited overshoot, but they do not reach zero globally.

Methane emissions

Black carbon emissions

Nitrous oxide emissions

FIGURE 10.3 Mitigation pathways to achieve the 1.5 degree global target

Source: IPCC, 2018

From a global perspective, the IPCC (2013, 2014a, 2014b) reports that they found high increases in heavy precipitation events, while droughts have become more frequent since the 1970s, especially in the (sub)tropics. Changes in large-scale atmospheric circulation and increases in tropical cyclone activity have also been documented since the 1970s (IPCC, 2013, 2014a, 2014b). The IPCC's latest Fifth Assessment Report highlights the observed and partly irreversible changes to the earth's ecosystems, particularly the changes to the oceans, that absorb a large part of the carbon dioxide and thereby become acidified, and the cryosphere (IPCC, 2013, 2014a, 2014b).

Another major change in the climate is related to sea-level rise. The IPCC reports that sea-level rise occurs for two main reasons:

*   thermal expansion of the oceans
*   discharge of additional water into the oceans as terrestrial ice and snow melt,

So far, the oceans have absorbed more than 80% of the heat added to the climate system (IPCC, 2007, 2014a, 2014b). Because of both thermal expansion of the ocean and meltwater discharge, the global average sea level has been rising. The IPCC's Fifth Assessment Report reports an average rate of sea-level rise of about 1.7mm (1.5–1.9mm) per year between 1901 and 2010, about 2.0mm (1.7–2.3mm) per year between 1971 and 2010, and about 3.2mm (2.8–3.6mm) per year between 1993 and 2010 (IPCC, 2013, 2014a, 2014b). A further rise in sea level is expected in the future.

As for the cryosphere (the part of the earth's surface covered by ice), the IPCC's Fifth Assessment Report mentions that over the last 20 years the Greenland ice sheet, the Antarctic ice sheet, glaciers worldwide and Arctic sea ice have declined. The IPCC gives an observed average rate of ice loss from glaciers worldwide of about 226 Gt per year over the period 1971–2009 and 275 Gt per year over the period 1993–2009 (IPCC, 2014a, 2014b). A further melting of ice sheets, glaciers and sea ice is expected in the future. More recently, scientists suggested that in fact the rate of sea-level rise might be underestimated due to a more rapid observed melting of the cryosphere than earlier anticipated. Even more striking changes in the cryosphere have been observed by several studies recently, such as a much higher rate of melting of glaciers, ice caps and ice sheets, particularly in the Arctic and Antarctic than previously anticipated. This also creates new geopolitical, environmental and socio-economic pressures as some countries are keen on exploiting minerals and energy resources in the Arctic (e.g. Russia and the USA).

> Also, the IPCC suggests that climate change can exacerbate disaster risks such as risks associated with droughts, the frequency and severity of tropical cyclones and flooding (IPCC, 2007, 2014a, 2014b). The United Nation's International Strategy for Disaster Reduction (ISDR) establishes a link between climate-related hazards and disasters. It reports that between 1988 and 2007 about two-thirds of all disaster events were of climatological nature. These climatological disaster events accounted for approximately 45% of the deaths and 80% of the economic losses caused by natural hazards. ISDR reports changes in long-term observed precipitation trends during 1990–2005 with more dry spells in the Sahel and the Southern African region, throughout the Mediterranean and parts of South Asia, increases in heat waves in many world regions and increases in heavy precipitation events. ISDR also reports an observed increase in the frequency and severity of droughts over wide regions since the 1970s, especially in the tropics and subtropics. It has also observed an increase in tropical cyclones, particularly over the North Atlantic, and coastal erosion due to intensified storms. Van Aalst assessed the relationship between climate change and extreme weather events and argues that potential increases in extreme events due to climate change are often exacerbated by increased socioeconomic vulnerability, thereby making the impacts even worse (Van Aalst, 2006).
>
> *(Urban et al., 2011: 11)*

These findings support the IPCC findings (e.g. IPCC, 2014a, 2014b, 2018). A further rise of climate-related hazards and extreme weather events is expected in the future.

Considering future impacts, the IPCC Fifth Assessment Report mentions that most climate scenarios assume an average surface temperature increase of more than 2 °C until 2100 and warming is likely to increase further in the 22nd century (IPCC, 2014a, 2014b). Figure 10.4 shows some projected climate impacts.

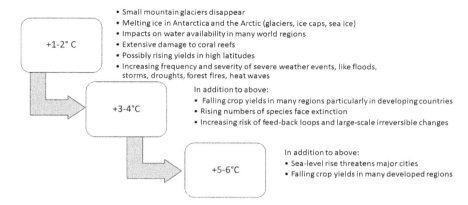

FIGURE 10.4   Observed and projected impacts of climate change, averaged global temperature increases

Source: Based on IPCC (2018)

Climate change policy

Climate change policy

A short history of the UN climate change negotiations

The United Nations Framework Convention on Climate Change (UNFCCC) developed an international climate treaty at the Earth Summit in Rio de Janeiro in 1992 and thereby acknowledged the role of humans in contributing to climate change. The treaty aimed to prevent dangerous climate change but made no commitment to emission reductions (UNFCCC, 1992; Urban & Nordensvärd, 2013).

This was followed by the Kyoto Protocol in 1997, which aimed to reduce GHG emissions to avoid dangerous climate change and had binding emission reduction commitments for developed countries for the first commitment period of the Kyoto Protocol for 2008–2012 (Urban & Nordensvärd, 2013). The second commitment period of the Kyoto Protocol was negotiated at the climate conference in Durban in late 2011 and covers the period from 2013 to 2020. Due to Article 10 of the Kyoto Protocol and the recognition of 'common but differentiated responsibilities' between developed and developing countries' historic responsibility for climate change, no emission reduction commitments for developing countries were embedded in the Kyoto Protocol (UNFCCC, 1998: 9; Urban & Nordensvärd, 2013).

Climate change mitigation also has a prominent role in the Bali Action Plan and the Bali Roadmap of 2007, and energy is central to these issues. Mitigation, hereby meaning reducing greenhouse gas emissions that lead to climate change and enhancing carbon sinks, is considered one of the five pillars of the Bali Action Plan (Urban & Nordensvärd, 2013). The aim of the Bali Action Plan and the Roadmap is to develop a strategy for a post–2012 climate agreement, for the time when the first commitment period of the Kyoto Protocol is over (Urban & Nordensvärd, 2013). In subsequent years, climate change mitigation took a prominent position in the UN climate change negotiations, and targets to mitigate emissions are one of the key elements over which major differences exist between developed and developing countries, but also within the group of developed and developing countries (Urban & Nordensvärd, 2013). This was apparent for the Copenhagen Accord in 2009, when no binding agreement could be reached on emission reduction and only the 'strong political will to combat climate change' was mentioned (UNFCCC, 2009: 5). This was also apparent in the Cancun Agreements in 2010, where major advances had been made in areas such as climate finance, technology transfer and Reducing Emissions for Deforestation and Forest Degradation (REDD+) (Urban & Nordensvärd, 2013). However, only a pledge of 'reducing global greenhouse gas emissions so as to hold the increase in global average temperature below 2 °C above preindustrial levels' was made (UNFCCC, 2010: 3). The climate change conference in Durban in late 2011 delivered further evidence on just how hard it is to agree on binding emission reduction targets for mitigating global climate change; it was agreed to postpone any legally binding global agreement – beyond those obliged by the Kyoto Protocol – until at least 2015, with implementation by 2020 (Urban & Nordensvärd, 2013).

There was a standstill in international climate negotiations for several years. Finally, in Paris at COP 21 in late 2015, there was a major breakthrough in the international climate change negotiations when the Paris Agreement was negotiated.

The Paris Agreement on climate change

The UNFCCC Paris Agreement on climate change is a global treaty that has been signed by 195 countries and came into effect in November 2016. The Paris Agreement's implementation starts in 2020 and has the following aims:

- To keep the global average air temperature to well below 2 °C above pre-industrial levels by 2100, with the ambition to keep temperature increases to 1.5 °C above pre-industrial levels to limit the risks and impacts of climate change. This also requires peaking global emissions as soon as possible.
- To promote adaptation to climate change (how to cope and live with climatic changes) to achieve climate-resilient development and low GHG emissions development.
- To raise climate finance for enabling poorer countries to adapt to climate change and reduce emissions. The required funding is US$100 billion per year by 2020 (UNFCCC, 2015).

The Paris Agreement is the first global climate agreement that encompasses all countries. In the Kyoto Protocol, there was a differentiation between developed countries (so-called Annex I countries) and developing countries (so-called non-Annex I countries), which divided the international climate community into those with emission reduction commitments (Annex I) and those without (non-Annex I). Under the Paris Agreement, all countries are obliged to take on emission reduction targets which are in line with their national development goals. These are called Nationally Determined Contributions (NDC). Hence, each country can submit a plan for mitigation action that is suitable for their own country and in line with its national priorities and needs. While high-income countries usually plan for deep emission cuts, such as the countries of the European Union, low- and middle-income countries usually aim to achieve mitigation and adaptation actions that foster their development needs at the same time, such as increasing energy access through renewable energy technology or protecting carbon sinks through afforestation and forest conservation. Many low- and middle-income countries have marginal emission reduction targets, as their emissions are minimal anyway. Others have very ambitious mitigation plan, such as those by Costa Rica that aims for carbon neutrality. While it was a major global diplomatic breakthrough to negotiate the Paris Agreement, political opposition to the agreement soon arose from the US under Trump's leadership. Trump has threatened for a number of years that the USA will leave the Paris Agreement. How this will work out practically and what the implications are for other climate-sceptic governments such as in Saudi Arabia, Russia and other OPEC countries remains to be seen.

Achieving the 1.5–2 C target

The Paris Agreement, as well as preceding climate policies such as the Copenhagen Accord, mentioned the below 2 C target to keep the global average temperature to well below 2 °C above pre-industrial levels by 2100. The 2 C target is regarded by scientists as the threshold between 'acceptable' and 'dangerous' climate change. A rise above 2 C by 2100 is likely to lead to abrupt and irreversible changes (IPCC, 2013). These changes may result in huge societal, economic and environmental disruptions, which are likely to threaten international development throughout the 21st century and beyond (Urban & Nordensvärd, 2013).

The IPCC and the wider scientific community overwhelmingly agree that climate change is caused by humans (so-called anthropogenic climate change), and there is agreement about the main climatic impacts and their severity. Yet, there is some uncertainty about the possibility of mitigating climate change, how fast (or how slow) mitigation is possible, how effective (or how ineffective) global and national responses will be, how economic, social, political and technological factors may speed up or impede mitigation and whether the 2 C target can be reached by the end of this century. Usually, future climate scenarios are used to simulate what will happen in the future under certain conditions. High, low and medium scenarios will be applied to take into account different variables such as population growth, economic performance, energy use, policies, technologies, etc. This results in a range of likely outcomes.

For many years, the scientific community was critical of achieving the 2 C targets, let alone the 1.5 C target, despite it being deemed possible at the political level, as can be witnessed in the international climate change negotiations that resulted in the Paris Agreement. More recently, a series of interesting studies by well-known scholars have been published that argue that the 1.5 C target is indeed still achievable (Millar et al., 2017), whereas others argue that it is unlikely (Raftery et al., 2017). Some may argue that achieving the 2 C target, and even more so the 1.5 C target by 2100, would depend strongly on advanced technology, including controversial technology such as carbon capture and storage (CCS) and/or bioenergy with carbon capture and storage (BECCS). BECCS is a future energy technology that produces negative carbon emissions. This requires energy to be produced from biomass, such as trees and crops that absorb carbon dioxide during their lifetime, then using this biomass to generate electricity or heat, and using CCS to inject the carbon dioxide back into geological deposits. Some critics argue this is a technical fix that is far away from today's technological reality, whereas others argue that this technology will be commercially available on a large scale well before 2100.

Exercises

1   Read the summary for policymakers of the Intergovernmental Panel on Climate Change (IPCC)'s report on limiting global atmospheric temperature increases to 1.5 degrees: www.ipcc.ch/sr15/chapter/summary-for-policy-makers/
    Make a list to note the following: (a) What climatic impacts have already been observed since pre-industrial times? What are the projected impacts of future climate change until 2100? (b) What are possible emission reduction options? What role does energy play for emission reduction pathways? (c) How can the 1.5 degree target be combined with wider sustainable development targets?
2   Imagine you were a policymaker for a country of your choice. What would you be suggesting to the parliament so that CO_2 emissions from energy-related activity would be reduced? Make a list of the top five strategies and present them in the form of a short speech.
3   Visit the NASA website on vital signs of observed climate change: https://climate.nasa.gov/vital-signs. Look at the time series model for CO_2 emissions and global temperature. What changes have been observed? Which regions have been effected disproportionally heavily by temperature increases? How will these changes affect our future climate? What can be done to avoid dangerous climate change in the future?

References

Barry, R.G. & Chorley, R.J. (2003) *Atmosphere, Weather and Climate*. 8th edition. London, Routledge.
Houghton, J. (2009) *Global Warming: The Complete Briefing*. 4th edition. Cambridge, Cambridge University Press. pp. 20–111.

IEA. (2019) *Statistics*. Paris, International Energy Agency (IEA), OECD/IEA. Available from: www.iea.org/statistics/

IPCC. (2007) *Climate Change 2007*. Synthesis Report. Contribution of Working Groups I, II and III to the Fourth Assessment Report of the Intergovernmental Panel on Climate Change (IPCC). [Core Writing Team, Pachauri, R.K and Reisinger, A. (Eds.)]. IPCC, Geneva, Switzerland, 104 pp. Available from: www.ipcc.ch/publications_and_data/publications_ipcc_fourth_ assessment_report_synthesis_report.htm

IPCC. (2013) *Climate Change 2013. The Physical Science Basis*. Contribution of Working Group I to the Fifth Assessment Report of the Intergovernmental Panel on Climate Change (IPCC). [Stocker, T.F., D. Qin, G.-K. Plattner, M. Tignor, S.K. Allen, J. Boschung, A. Nauels, Y. Xia, V. Bex & P.M. Midgley (Eds.)]. Cambridge University Press, Cambridge, United Kingdom and New York, NY, USA, 1535 pp. Available from: www.climatechange2013.org/

IPCC. (2014a) *Climate Change 2014. Impacts, Adaptation and Vulnerability*. Contribution of Working Group II to the Fifth Assessment Report of the Intergovernmental Panel on Climate Change (IPCC). [Field, C.B., V.R. Barros, D.J. Dokken, K.J. Mach, M.D. Mastrandrea, T.E. Bilir, M. Chatterjee, K.L. Ebi, Y.O. Estrada, R.C. Genova, B. Girma, E.S. Kissel, A.N. Levy, S. MacCracken, P.R. Mastrandrea, and L.L. White (Eds.)]. Cambridge University Press, Cambridge, United Kingdom and New York, NY, USA. Available from: www.ipcc.ch/report/ar5/wg2/

IPCC. (2014b) *Climate Change 2014. Mitigation of Climate Change*. Contribution of Working Group III to the Fifth Assessment Report of the Intergovernmental Panel on Climate Change (IPCC). [Edenhofer, O., R. Pichs-Madruga, Y. Sokona, E. Farahani, S. Kadner, K. Seyboth, A. Adler, I. Baum, S. Brunner, P. Eickemeier, B. Kriemann, J. Savolainen, S. Schlömer, C. von Stechow, T. Zwickel and J.C. Minx (Eds.)]. Cambridge University Press, Cambridge, United Kingdom and New York, NY, USA. Available from: www.ipcc.ch/report/ar5/wg3/

IPCC (2018) *Global Warming of 1.5 degrees*. Available from: www.ipcc.ch/site/assets/uploads/sites/2/2018/07/SR15_SPM_version_stand_alone_LR.pdf

Millar, R.J., Fuglestvedt, J.S., Friedlingstein, P., Rogelj, J., Grubb, M.J., Matthews, H.D., Skeie, R.B., Forster, P.M., Frame, D.J. & Allen, M.R. (2017) Emission budgets and pathways consistent with limiting warming to 1.5 °C. *Nature Geoscience*, 10, 741–747.

NASA. 2019. *Carbon Dioxide Measurements*. Available from: https://climate.nasa.gov/

Quadrelli, R. & Peterson, S. (2007) The energy – climate challenge: Recent trends in CO_2 emissions from fuel combustion. *Energy Policy*, 35 (11), 5938–5952.

Raftery, A.E., Zimmer, A., Frierson, D.M.W., Startz, R. & Liu, P. (2017) Less than 2 °C warming by 2100 unlikely. *Nature Climate Change*, 7, 637–641.

Richardson, K., Steffen, W., Schellnhuber, H.J., Alcamo, J., Barker, T., Kammen, D.M., Leemans, R., Liverman, D., Munasinghe, M., Osman-Elasha, B., Stern, N. & Wæver, O. (2009) *Synthesis Report. Climate Change. Global Risks, Challenges and Decisions*. 10–12 March 2009, Copenhagen. University of Copenhagen. Available from: www.pik-potsdam.de /news/press-releases/files/synthesis-report-web.pdf

UNFCCC. (1998) *Fact Sheet: The Need for Mitigation*. Available from: https://unfccc.int/files/press/backgrounders/application/pdf/press_factsh_mitigation.pdf

UNFCCC. (2009) *Copenhagen Accord*. Available from: https://unfccc.int/resource/docs/2009/cop15/eng/11a01.pdf

UNFCCC. (2010) *Cancun Agreements*. Available from: https://unfccc.int/tools/cancun/index.html

UNFCCC. (2015) *Paris Agreement*. Available from: https://unfccc.int/sites/default/files/english_paris_agreement.pdf

Urban, F. & Nordensvärd, J. (2013). *Low Carbon World Coal Extraction Has Very Recently Peaked and Many Countries are Reducing Their Use of Coal for Electricity Development: Key Issues.* Abingdon, Earthscan, Routledge.

Urban, F. & Mitchell, T. (2011). *Climate Change, Disasters and the Energy Sector: Issues at the Interface of Changing Disaster Risks, Adaptation and Mitigation. SCR Paper.* Brighton, IDS.

Van Aalst, M. (2006). The impacts of climate change on the risk of natural disasters. *Disasters*, 30(1), 5–18. https://doi.org/10.1111/j.1467-9523.2006.00303.x

11

ENVIRONMENTAL IMPLICATIONS – NATURAL RESOURCE DEPLETION AND AIR POLLUTION

Natural resource depletion and degradation

Fossil fuel depletion

A key environmental impact of energy use is natural resource depletion. This section will address the environmental implications of fossil fuel depletion. Fossil fuel resources have been formed over millions of years from the organic remains of prehistoric animals and plants. They have a high carbon content and include coal, oil and natural gas. These fossil fuels are non–renewable energy sources, as their reserves are being depleted much faster than new ones are being formed. For example, it took millions of years to form an oil field, but it can take only a few years of exploitation to deplete it. Mining and extraction are therefore major causes of the depletion of fossil fuel resources. This leads to severe environmental impacts, including the destruction of natural ecosystems (Goldemberg & Lucon, 2009).

Coal mining and oil field exploration are the most environmentally unfriendly forms of fossil fuel resource depletion. Coal mining involves the destruction and degradation of natural ecosystems and landscapes, both underground and above the ground. This means a loss of habitat for animals and plants and threats to biodiversity. Oil field exploration used to be located onshore but is now often located offshore too. Like coal mining, it involves the destruction and degradation of natural ecosystems, landscapes and habitats for animals and plants, often including sensitive marine ecosystems. Oil exploration also comes with the risk of polluting rivers, oceans, groundwater and other water bodies. This can happen, for example, when a leak occurs while drilling for oil. We will discuss these issues further when we discuss water pollution from energy sources later in this chapter.

A key issue related to energy use and natural resource depletion is peak oil. Peak oil is a concept that describes first an increase in oil production up to a peak and afterwards a decline in oil production. This is based on an observed rise of oil production, a peak and then a fall in the production rate of oil fields over time as the oil resources are depleted. The theory is that this phenomenon of peaking oil production not only is limited to oil fields, but also applies globally as all oil resources could be depleted at some point. This is due to rapid oil extraction from finite natural resources that can be depleted and that needed millions of years to be built. Peak oil is therefore the point of maximum oil production. There is a lively debate among scientists and the oil lobby as to whether peak oil has already happened or whether it still lies ahead. Some experts argue that we are already beyond peak oil as the rate of depletion is rapid, oil prices have increased to formerly unseen levels, few new oil resources are being found and the extraction of unconventional oil resources, such as from tar sands and shale oil, has expanded rapidly in recent years. This extraction of unconventional oil resources involves environmentally destructive methods such as fracking (Leggett, 2013). The UK Energy Research Centre published a systematic review report that reviewed over 500 studies on peak oil and global supply forecasts and concluded that a peak in oil production is likely to happen before 2030, if not earlier (UK ERC, 2009). Critics argue that new resources are being found which will add to the supply of fossil fuel resources. However, fossil fuel resources formed over millions of years, and humans are currently exploiting them within a few decades. This means that even when new fossil fuel resources are being found frequently, extraction rates are too fast to avoid depletion of these resources.

According to calculations by Roper (2019), world coal extraction has very recently peaked and many countries are reducing their use of coal for electricity generation. There might be a potential second peak around 2050–2060, mainly based on the demands for coal in China and other newly industrialised economies.

Roper (2019) further argues that world oil extraction may be likely to peak between 2020 and 2030, after which a smaller amount of oil will be extracted as reserves could run low, even when unconventional oil resources such as tight crude oil are taken into account (see Figure 11.1). For both coal and oil, it is interesting to see that experts such as Roper, who bases his calculations on figures from the US Energy Information Administration (EIA), predict a decline in the extraction rate (and therefore the available fossil resources) within the next few years or the next decade at the latest. We can see from Figures 11.1. and 11.2 that the assumption is that peak oil will happen very soon. This is in line with the report of the UK Energy Research Centre (UK ERC, 2009).

The projections for world natural gas depletion are, however, more optimistic. Roper (2019) argues that natural gas extraction could peak later, at around 2030–2035, due to an increase in unconventional natural gas extraction, such as shale gas (Figure 11.2). Nevertheless, it is estimated that even with the exploitation of unconventional natural gas, production may decline rapidly after 2035 due to a depletion of reserves.

The depletion of fossil fuel resources also has implications for energy security. Energy security is defined as availability of affordable energy supply at adequate times, in adequate quantities and adequate quality to an extent that guarantees the

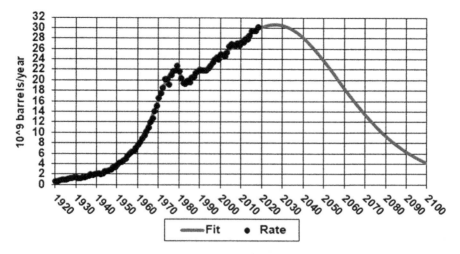

FIGURE 11.1   World oil extraction: global peak

Source: Roper (2019)

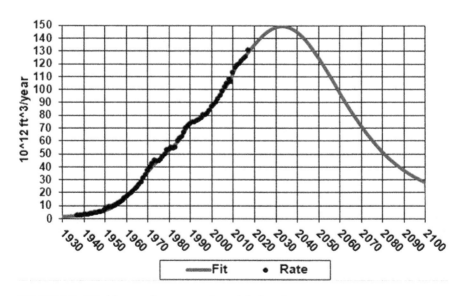

FIGURE 11.2   World natural gas extraction: global peak

Source: Roper (2019)

economic and social development of a country (Kowalski & Vilogorac, 2008). As not every country has abundant fossil fuel resources, they are dependent on imports from the Organization of the Petroleum Exporting Countries (OPEC). While some of the OPEC countries are relatively stable, others are politically unstable and volatile. Depending on imports from these countries may be a risk to national energy security and could jeopardise economic growth and prosperity. Relying on more domestic low carbon energy sources, such as renewable energy, can therefore help to increase national energy security and reduce risky and expensive imports.

Deforestation

After discussing how energy use and fossil fuel resource depletion are linked, we discuss how energy use and deforestation are linked. Traditional biomass use such as fuelwood use and charcoal use can be directly linked to deforestation.

Traditional biofuels are primarily used domestically for cooking and heating, particularly in poorer rural areas, with energy poverty hotspots in Sub-Saharan Africa and developing Asia. Consumption of traditional biofuels is mainly due to three factors:

a   income levels – higher levels may mean lower use of traditional biofuels due to being able to afford other energy sources
b   the degree of urbanisation – higher levels may mean more availability of alternative energy sources, hence lower use of traditional biofuels
c   the degree of industrialisation – higher levels may mean lower use of traditional biofuels as energy systems may be more advanced

The share of traditional biofuels declines in general with rising gross domestic product (GDP), although differences exist between countries and regions (Victor & Victor, 2002). When income distribution and geographic distribution is taken into account, the relation between income and traditional biofuel use, particularly fuelwood use, can be different (Victor & Victor, 2002), because traditional biofuels are mainly used by lower-income households and in poorer rural areas (Urban, 2014).

However, many poor countries suffer not only a lack of access to modern energy, but also a lack of access to energy markets, services and infrastructure. This is particularly distinct in rural areas that are often far away from the grid, from cities, trading centres and financial and technological services. As alternative energy sources may not be available, people often resort to using fuelwood. The harvesting and use of traditional biomass can in some cases lead to the depletion of natural forest resources. While many people rely on the collection of fallen branches from trees and non-forest fuelwood sources, others depend on felling trees. The production of charcoal also involves the felling of trees. This can lead to a decrease of woodlands and forests that can in some localised cases and under certain conditions lead to larger-scale deforestation, erosion and desertification. This practice can also lead to a decrease in biodiversity and negative impacts on water and food security. In some areas where forests offer some protection against natural disasters, felling trees for fuelwood can have devastating impacts. For example, as mangrove forests are cleared for fuelwood, this reduces the natural protection from floods, tsunamis and sea-level rise, thereby leaving people, economies and ecosystems even more vulnerable to natural disasters and climatic impacts.

In the 1970s, theories emerged about a fuelwood gap. This was based on the assumption that fuelwood collection leads to large-scale deforestation and that there would not be enough forest supply available to meet the growing demand for fuelwood, hence the gap in fuelwood supply. In the 1990s, the fuelwood gap theory was rejected as it was found that large-scale deforestation had not occurred due to fuelwood collection. It rather emerged that other factors were the main drivers of large-scale deforestation, for example, land use and land conversions (e.g. deforestation to transform forests into agricultural land, to extract natural resources or for building sites). Population growth also plays a role in natural resource depletion. It has emerged that there is a high degree of uncertainty associated with assumptions about fuelwood use as the data tend to be rather inaccurate and incomplete (FAO, 1997). Studies by Fairhead and Leach (1998) suggested that local communities were often falsely accused of deforestation due to fuelwood use and agricultural practices. Their research found that in West Africa, contrary to common perceptions, local communities actually contributed to afforestation, rather than deforestation, due to their local practices.

While many of the findings of the 1990s are still valid today, new energy-related pressures have emerged, such as deforestation and land grabs (or 'green grabs' as described by Fairhead et al., 2012) for biofuel production, rather than for fuelwood.

Water

Water and energy are inherently linked. Water is needed for fossil fuel extraction, particularly for unconventional fossil fuels, such as for fracking. As the name hydraulic fracturing, commonly known as fracking, suggests, this method of extracting unconventional fossil fuels such as shale gas, shale oil, tight gas, tight oil and coal-bed methane requires water. Large amounts of water, sand and chemicals are injected at high pressure into boreholes in deep rock formations. This mining technique results in small fractures in the rock formations, which enable the gas and oil products in the fractures to be exploited. This procedure tends to be expensive and requires a lot of energy and water. Millions of tonnes of freshwater are needed for fracking one shale gas well. There is concern that fracking is increasing water stress in areas that are already drought prone, such as arid areas in Texas, Colorado, Utah, New Mexico and California in the USA. This has raised concerns about depleting drinking water resources and the negative impacts on aquatic environments (Mielke et al., 2010). Some firms are recycling wastewater; however, there are concerns that the amounts of water recycled are too little to offset the large amounts of water needed for fracking. There are also concerns about the potential contamination of groundwater due to the use of chemicals that are needed alongside water for fracking, as well as the release of air pollutants and greenhouse gas (GHG) emissions that contribute to climate change. There are also health concerns associated with fracking (Kovats et al., 2014). Fracking is likely to contribute to outdoor air pollution through the release of volatile organic compounds, tropospheric ozone, and diesel particulate matter. Pollutants are used that can contaminate ground water and surface water, such as benzene, hydrocarbons, endocrine-disrupting chemicals and heavy metals (Kovats et al., 2014).Some countries, such as France and recently Germany, have banned fracking; others are promoting fracking, such as the USA and the UK.

Even for conventional fossil fuels, large amounts of water are used for mining, processing and operating power plants. Coal mining, for example, requires water for coal extraction, mining, washing and processing. Water is also needed for combusting coal in power plants, for example, for generating steam and for cooling generating systems. It is estimated that in China, a fifth of all water used domestically is due to coal mining, processing and power plant use (Odgaard, 2014).

At the same time, water is embedded within products and services. This refers to the notion of 'embedded water' or 'virtual water' and is linked to the trade of virtual water (Dabrowski et al., 2009; Aldaya et al., 2010). Examples are the virtual water that is embedded in clothes, such as cotton shirts, or meat products, such as hamburgers. This concept also applies to energy as the virtual water that is embedded in power plants, nuclear reactors, wind turbines, solar panels, grids and electricity generation. While the concept of virtual water has been criticised by some experts as not offering sufficient insight for policymakers (Wichelns, 2010a, 2010b), experts agree that it is an important tool to assess the impact of humans on scarce water resources. A measure to assess the impact of human lifestyle and consumption on natural resources is the water footprint. 'The water footprint of a country is defined as the volume of water needed for the production of the goods and services consumed by the inhabitants of the country' (Hoekstra & Chapagain, 2007: 35). Agricultural products, particularly crops and cereals, and animal products, such as meat and dairy products, make up the largest share of the global water footprint (Hoekstra & Mekonnen, 2012).

Four factors affect the water footprint of a country (Hoekstra & Chapagain, 2007):

- the volume of water consumption
- the consumption pattern, such as high or low meat consumption
- the climate and growth conditions
- agricultural practices, such as water use efficiency

So far, the virtual water of energy extraction, generation and use has received limited attention; however, this is changing as unsustainable and water-consuming energy practices, such as fracking, generate more criticism.

Another major impact of energy extraction and generation on water resources is through water pollution from oil resources. Three prominent examples of water pollution from oil exploitation are as follows:

- The pollution of oceans and marine environments from oil exploitation activities on oil rigs, such as the oil disaster that happened at the BP Deepwater Horizon in the Gulf of Mexico off the shores of the USA in 2010. The oil rig exploded, releasing oil pollution of several tens of thousands of barrels per day. It is classified as one of the world's worst environmental catastrophes. BP had to pay several billion US dollars in compensation for violating the US Clean Water Act.

- The pollution of oceans and marine environments from shipping, such as the oil spills of the oil tankers Amoco Cadiz in 1978 off the coast of France, the Exxon Valdez in 1989 in Alaskan waters and the Prestige in 2002 off the coast of Spain. Hundreds of thousands of barrels of crude oil spilled from these tankers into fragile marine environments, polluting the seawater, marine fauna and flora, shores and beaches with highly toxic substances.
- The pollution of freshwater and groundwater from oil spills, such as the polluting oil industry in the Niger Delta in Nigeria, is caused mainly by the poor performance and lack of environmental protection of oil industries such as Shell, Texaco and Exxon Mobil, and joint ventures with the Nigerian state. A smaller percentage of the oil spills also occurs due to sabotage and theft. Estimates suggest that many tens of millions of barrels of oil have polluted the environment since the 1960s. Oil spills have led to the pollution of seawater, groundwater and drinking water, fish and other fauna and flora, declines in biodiversity, loss of habitable land, loss of arable land, vegetation and ecosystems (such as islands, mangrove ecosystems, swamps and rainforests). In addition, oil pollution has severe health impacts. Natural gas flaring also leads to environmental pollution and the release of toxic substances.

(Nwilo & Badejo, 2008)

Water is also needed for hydropower. While small-scale hydropower has fewer effects on the environment and on people, large hydropower dams involve a range of water-related environmental impacts. Dams can be built for energy generation and for flood control, which are, in principle, two positive attributes of dams. However, large dams and reservoirs may come with a series of environmental impacts, such as a decline in water quality, changes in river flow and hydrology with impacts both upstream and downstream, changes to sediment loads, eutrophication, impacts on aquatic fauna and flora, geomorphological changes, risk of landslides, etc. There are also debates about dam-induced seismicity, which might be one factor triggering earthquakes, for example, in China and elsewhere where dams are built on tectonic fault lines. There are also emissions of GHGs from reservoirs; for example, methane from the decay of flooded biomass. Nevertheless, these GHG emissions are far below those of fossil fuel plants (Goldemberg & Lucon, 2009).

Land

The water–food–energy nexus is often closely linked to land. Cooke et al. (2015) suggest that land is often the starting point for energy generation and production. Land is needed for coal mining and oil and natural gas extraction. Oil exploration is particularly polluting and can pollute soils and land resources, as in Nigeria. This in turn has impacts on agriculture and food security, as well as on groundwater and freshwater resources, as discussed previously.

Land is also needed for hydropower developments, such as dams, and for large-scale renewable energy, such as large wind and solar PV farms. Cooke et al. (2015) use the case of the Bakun mega-dam in Sarawak, Borneo, Malaysia, to exemplify how land rights and land grabs are linked to energy development. In the case of the Bakun dam, indigenous peoples of Sarawak were evicted from their land (as they did not own the land, but lived on it for many centuries). They were relocated into resettlement housing due to the construction of the large dam, housing which was regarded as too small or sub-standard by many residents. The Bakun dam provides electricity for industrial activities and urban centres in Sarawak. While the benefits of the dam are being felt far away, such as in Sarawak's capital Kuching, the local impacts are being felt by indigenous groups close to the dam site. The dam involved flooding old-growth rainforest that is rich in biodiversity and hosts a number of endangered species, such as the orang-utan. The dam destroyed and degraded the natural tropical forest ecosystem, cut into the habitat of rare, endemic and endangered species, reduced biodiversity and gave rise to a number of negative environmental impacts such as soil erosion, decline in water quality, changes in river flow with impacts both upstream and downstream, changes to sediment loads, eutrophication, impacts on aquatic fauna and flora, geomorphological changes, etc. The indigenous people lost access to their land, forest and the river. Similar conclusions have been drawn from dams in Brazil (Goldemberg & Lucon, 2009). Land, and a wide range of land-rights issues, are therefore at the heart of the debate about water, food and energy security.

Interestingly, dams are not the only problem associated with land in Borneo. While Borneo is known as one of the world's last great wildernesses, it is far from being untouched. Deforestation and forest degradation for monoculture plantations have existed in Borneo since it was first colonised. The British colonisers deforested parts of the old-growth tropical rainforest in Borneo to create monoculture plantations of tobacco and rubber plants. More recently, domestic Malaysian and Indonesian (and some foreign) firms have exploited the land and the forest to cultivate oil palm. The oil palm plantations produce plant-based oil for a wide range of products, ranging from food products to beauty products to biofuels. Biofuels for low carbon transport are therefore placing land at the heart of the debate about deforestation, as well as the discussion around land for food versus land for fuel. We will briefly review the biofuel issue.

First-generation or conventional biofuels are usually based on edible biomass-based starch, sugar or vegetable oil. This means these fuels are usually based on food products, such as corn, wheat or other cereals, cassava, sugar beets, sugar cane, that are used for making bioethanol. Soy, jatropha and oil palm are used for making biodiesel. Second-generation biofuels are usually made from feedstock and waste (e.g. municipal waste). Third-generation biofuels or advanced/unconventional biofuels usually do not depend on food products or feedstock, but can be derived from algae, cellulose and other forms of plant biomass, which makes it harder to extract fuel (Goldemberg & Lucon, 2009). Due to their compatibility with food security, third-generation biofuels should therefore be favoured and investments in R&D should be promoted.

Many countries are increasingly producing first- and second-generation biofuels for transport. Unfortunately, it has been reported that this may negatively affect food security. The food security issue arises because land used for biofuel production cannot be used for food production; hence, there is a conflict between land for biofuels and land for food production. To make things worse, some biofuel crops, such as sweet potato or cassava, are also food crops (Rathmann et al., 2010). At the same time, there are allegations that some biofuel operations, including those of wealthy corporations (and often supported or at least tolerated by government agencies), may have evicted poor people in developing countries from their lands to gain access to land for growing biofuels. Hence, some biofuel developments are reported to potentially be associated with land grabs (Neville & Dauvergne, 2012; Oxfam, 2012).

Other forms of environmental degradation

Other forms of environmental impacts from energy use include the degradation and loss of ecosystems, biodiversity, habitat and the destruction of landscapes due to various forms of energy extraction, processing, generation, transport and transmission. For example, we discussed how oil spills from oil extraction, processing and transport, from oil tankers transporting crude oil and from oil rigs extracting oil, can lead to degradation and destruction of marine and terrestrial ecosystems, declines in biodiversity and loss of habitats for fauna and flora. Another example is large dams that can alter how the landscape looks and a large number of wind farms or solar PV plants in close proximity that can lead to landscape fragmentation.

Another environmental impact is the environmental pollution caused by uranium mining (and the mining of other nuclear materials) and the radioactive waste from nuclear power plants. Radioactive waste, such as uranium and plutonium, contains carcinogenic substances, such as radioactive strontium, iodine, caesium and plutonium. The half-life ($t\frac{1}{2}$) of radioactive substances can be many millions to several billion years (US NRC, 2002). Long-life radioactive waste must therefore be contained and isolated from the environment and from humans for very long periods due to its radioactivity. A common practice is therefore to deposit radioactive waste in geological formations, such as salt stocks. Nevertheless, the waste will have to be deposited safely in these ultimate disposal places for many thousands of years and longer, which makes the long-term governance and security of the radioactive waste sites extremely challenging. If the radioactive waste was released, this could result in radioactive pollution of the environment and could lead to gene mutations and morbidity for humans, animals and plants. Two key controversies of nuclear energy were exemplified in the nuclear disasters in Chernobyl, Ukraine (part of the former USSR), in 1986 and in Fukushima, Japan, in 2011. Even today, more than 25 years after the nuclear accident in Chernobyl, the affected area – including its water, soil, flora and fauna – is heavily contaminated, which in turn has impacts on human health (WHO, 2005). Another problem with nuclear energy is that it depends mainly on uranium, which is a finite resource. Nuclear energy, therefore, falls into the category of non-fossil resources. There is heated debate whether uranium will become a rare and near-depleted resource any time soon, similar to oil, or whether uranium resources will remain abundant for hundreds or even thousands of years to come.

The issue of natural resource depletion is valid for every component of the process of energy extraction, processing, generation, transport, transmission and use. Natural resources are used, for example, when power plants or components of energy systems are built and manufactured. An environmental impact of solar photovoltaic (PV) systems is that it depends on silicon for manufacture. Although there is plenty of silicon available, as it is derived from sand at affordable prices, PV systems depend on the mining of this natural resource.

Another environmental impact of energy use is black carbon. Black carbon is a component of particulate matter (PM ≤ 2.5 μm, also called PM2.5) (see also the next section on air pollution). It is created through the incomplete combustion of fossil fuels, biofuels and biomass. It has health impacts, leading to human morbidity, and is also a climate-forcing agent that contributes to global warming.

There is a wide range of environmental implications of energy use, energy extraction, processing, generation, transport and transmission. It is widely acknowledged that fossil fuels are the energy sources with the most environmental implications by far. Nevertheless, low carbon energy sources, such as nuclear energy and hydropower dams, also have environmental impacts (Goldemberg & Lucon, 2009). Even small-scale renewable energy options, such as PV systems and wind farms, can have some environmental impact, although they are marginal compared with the effect of fossil fuels. In general, all energy sources have some trade-offs of an environmental (and sometimes also social) nature, although some of these are marginal. While energy is vital for humans and for powering societies and economies, energy remains to some extent a dirty business.

Air pollution

Local air pollution

The World Health Organization (WHO, 2016) suggests that:

> Air pollution is a major environmental risk to health. By reducing air pollution levels, countries can reduce the burden of disease from stroke, heart disease, lung cancer, and both chronic and acute respiratory diseases, including asthma.
>
> The lower the levels of air pollution, the better the cardiovascular and respiratory health of the population will be, both long- and short-term.
>
> *WHO (2016: 1)*

We discussed air pollution from a social perspective in Chapter 7, focussing particularly on the health impacts of air pollution on people. In this regard, we also discussed indoor air pollution, which is a form of air pollution that does little harm to the environment but significantly affects people's health and their lives. This section will discuss air pollution from the perspective of the environmental implications.

Air pollution is defined as follows:

> A condition of 'air pollution' may be defined as a situation in which substances that result from anthropogenic activities are present at concentrations sufficiently high above their normal ambient levels to produce a measurable effect on humans, animals, vegetation, or materials.
>
> *Seinfeld and Pandis (2006: 21)*

Local air pollution is an environmental impact that is created through the combustion of fossil fuels from transport, power generation and industry, as well as from households (e.g. heating). The main sources are coal and oil. It leads to the emissions of sulfur dioxide (SO_2), carbon monoxide (CO), nitrogen oxides (NO_x), particulate matter (PM) as well as evaporative emissions of hydrocarbons (HCs) and volatile organic compounds (VOCs). The formation of the pollutant ozone (O_3) is caused in low altitudes by a chemical reaction of NO_x and HCs triggered by solar radiation. Local air pollution is often present in urban areas (Goldemberg & Lucon, 2009).

Another problem associated with local air pollution is smog. The combination of smoke and fog is called smog and is common both in the winter in cold climates and in the summer in warm climates. Three cities that are famous for smog are London, particularly seen from a historical perspective as air pollution has become better in recent decades; Los Angeles, which suffers from smog on warm, sunny days; and Beijing, which suffers from smog both on cold winter days and warm, sunny days. The smog on warm, sunny days is caused by tropospheric ozone that undergoes a photochemical reaction in the atmosphere. From an environmental perspective, smog can cause cell damage to plants and thereby contributes to vegetation and ecosystem degradation (Smithson et al., 2008). The use of more modern fuels like electricity and natural gas has reduced sulfur dioxide emissions, and policies to reduce air pollution have also played an important role. While historically the 'policy of high chimneys' for polluting industries and a move towards modern energy enabled a reduction in air pollution in many OECD countries such as the UK, the recent problem in air pollution in many cities, such as Los Angeles and Beijing, is mainly caused by excessive traffic.

Many of the world's cities, such as Accra, Beijing, Bogota, Cairo, Delhi, Dhaka, Karachi, Kolkata, Lima, Mexico City, Mumbai, Shanghai, Los Angeles, London and Milano, have considerable air pollution problems. While the problem is more pronounced in developing countries, due to less stringent environmental regulation and policies, some cities in developed countries continue to suffer from urban air pollution, such as Los Angeles. Pollution measurements show how WHO's guidelines for PM10 and PM2.5, two main indicators of air quality and air pollution, are being exceeded several times by many of the world's megacities, most strikingly in Delhi, Kolkata and Beijing (Parrish, 2013). WHO (2016: 1) reports that the 'major components of PM are sulfate, nitrates, ammonia, sodium chloride, black carbon, mineral dust and water. It consists of a complex mixture of solid and liquid particles of organic and inorganic substances suspended in the air'. The smaller these particles are, the easier they can be breathed in and lodged in the lungs and then cause damage to health. Some of these chemicals also cause acid precipitation (e.g. sulfates, nitrates) and can form greenhouse gases when reacting with other atoms.

This situation is made worse by increasing private car ownership in many Asian cities, such as Beijing. Beijing has introduced several policies to restrict personal vehicle use, such as restricting vehicle use on certain week days based on the registration number, partly as a result of policies introduced for the 2008 Olympics (Cai & Xie, 2011; Hao et al., 2011). In 2013, Beijing's local government raised an alarm as WHO's standards for air pollution were dangerously exceeded. Local residents were encouraged to stay indoors and to leave their homes only in exceptional circumstances (BBC, 2013). Air pollution in Beijing has been bad for several decades, and there is a saying that one day in Beijing is as bad as smoking one pack of cigarettes. Nevertheless, efforts were made by government officials in the run-up to the 2008 Olympics to improve air quality by closing down old, inefficient coal-fired power plants, increasing the share of renewable energy and natural gas among the energy supply, and closing down polluting factories and relocating them to neighbouring provinces (Urban, 2014). Interestingly, a part of the pollution in Beijing today comes from Hebei province, where the relocated industries are based. Today many Indian cities top the league of the world's most polluted cities.

To overcome this problem in the long run, energy supplies need to include a greater share of non-fossil fuels, such as renewable energy and other low carbon energy, and the transport system needs to create a lower demand for private fossil-fuel-driven vehicles by offering more opportunities for public transport, cycling, walking and low emissions vehicles, such as electric vehicles.

Transboundary air pollution

One of the problems with long-range and transboundary air pollution from fossil fuel combustion is that the pollution can be caused far away from the affected area; for example, pollution sources in urban areas can cause adverse effects in rural areas and even cross boundaries, and thereby effect other countries. A striking example is the case of the German Democratic Republic (GDR) up until 1989; pollution from the fossil fuel industries in the GDR was blown over the border into Germany and affected the people in places such as Bavaria.

Local air pollution can have negative impacts on human health as well as on fauna and flora, for example, by having adverse effects on plant growth and functionality, animal health and ecosystems and their services. Local air pollution can lead to acid rain (or acid precipitation) caused by sulfur dioxide and nitrogen dioxide. This means that the rain, or other precipitation such as snow, has a lower, more acid pH than ordinary rain. It leads to a variety of ecological damage, such as the acidification of lakes and other water bodies. This leads to the death of fish and other aquatic organisms. The problem has been particularly pronounced in Scandinavia due to transboundary air pollution from the Soviet Union and Eastern Bloc countries during the 1970s to 1990s (Goldemberg & Lucon, 2009).

Another phenomenon that is related to acid rain is forest dieback and more generally the death of plants. Forest dieback was common in the 1980s and 1990s in West German regions that bordered the GDR, such as the Fichtelgebirge, a forest region in Eastern Bavaria. The trees died because of transboundary air pollution and acid rain from the polluting coal-fired industries, heating systems and power plants of neighbouring GDR. Acid rain also corrodes stones, particularly marble and limestone that are more sensitive to acid (Goldemberg & Lucon, 2009).

To tackle transboundary pollution, the UN Convention in Long-Range Transboundary Air Pollution came into force in 1983. Since climate change has taken the limelight with regard to environmental 'catastrophes', the media, public and even science gives little attention today to acid rain and related forest dieback. Another reason is that the fall of the Soviet Union and the Eastern Bloc countries meant that many of the polluting coal-fired industries and the outdated heating and transport systems were replaced by more modern energy generation facilities. As a positive result, there is now much less acid rain and much less forest dieback from acid rain than there was in the 1980s and 1990s. In addition, fewer water bodies are suffering from acidification today in OECD countries.

Figure 11.3 shows how acid rain occurs.

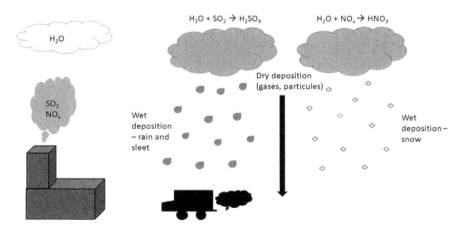

FIGURE 11.3 The main mechanisms of acid precipitation

Source: Author

Linking energy access, climate change and air pollution

There are synergies between energy access, air pollution, climate change and health. Positive synergies can open up opportunities for proving energy access by using renewable energy for electricity provision and for cooking fuels, such as biofuels. This would, in turn, reduce indoor and outdoor air pollution, as well as reduce greenhouse gas emissions that contribute to climate change. Negative synergies would mean a heavy reliance on fossil fuels for providing energy access, hence increasing air pollution and contributing to climate change.

Rao et al. (2013), Haines (2012) and Haines et al. (2010) point out that integrated governance is needed to develop a set of interlinked policies that address energy issues (including energy for households and for transport) as well as aim to reduce pollutants such as PM10 and PM2.5. Transport policies to address these synergies could be, for example, to increase the use, reach and efficiency of public transport (e.g. by introducing trams/metros/electric buses, extending the network, increasing the frequency of the service), introduce park and ride schemes, promote car sharing, improve road design to include bike lanes and pavements for pedestrians, provide dedicated parking and charging spaces for electric vehicles, which would all reduce dependency on petrol/diesel-driven private cars and thereby have positive effects on air pollution, climate change, natural resource use and health. Household policies to address these synergies could be, for example, to increase the use of clean, modern and affordable cooking fuels such as biogas or natural gas (rather than coal or charcoal), to introduce solar water heaters in place of electric/natural gas boilers, to promote the use of rooftop PV, and to introduce smart meters to monitor and reduce energy use within a household.

Exercises

1   Explain the link between energy generation/use and impacts on natural resources such as land, water, fossil reserves, etc.
2   Look at Figures 11.1, 11.2 and 11.3. What do these three graphs show concerning energy use and the availability of fossil fuel resources? What does this mean from an investment point of view for a country that is in need of investing in new energy generation infrastructure? Can you think of other alternatives for energy generation that are not related to limited resources?
3   Check out the real time air pollution map at https://waqi.info/. Which regions are currently heavily affected by air pollution? Which regions are relatively free from air pollution? How might this change over time as the day/night progresses (e.g. in relation to peak hours of traffic, working hours for industries and other important factors affecting air quality)?

References

Aldaya, M.M., Martínez-Santos, P. & Llamas, M.R. (2010) Incorporating the water footprint and virtual water into policy: Reflections from the Mancha occidental region, Spain. *Water Resources Management*, 24 (5), 941–958.

BBC (12 January 2013) *Beijing Air Pollution Soars to Hazard Level*. BBC World News, China. Available from: www.bbc.co.uk/news/world-asia-china-20998147

Cai, H. & Xie, S. (2011) Traffic-related air pollution modeling during the 2008 Beijing Olympic Games: The effects of an odd – even day traffic restriction scheme. *Science of the Total Environment*, 409 (10), 1935–1948.

Cooke, F.M., Saat, G.B., Gani, N. & Urban, F. (2015) The land issue in the water, energy and food security nexus. In Hezri, A., (Ed.), *Water, Energy and Food Security Nexus*. Singapore: Konrad Adenauer Foundation, pp. 123–131.

Dabrowski, J.M., Murray, K. & Ashton, P.J. (2009) Agricultural impacts on water quality and implications for virtual water trading decisions. *Ecological Economics*, 68 (4), 1074–1082.

Fairhead, J. & Leach, M. (1998) *Reframing Deforestation: Global Analysis and Local Realities: Studies in West Africa*. Global Environmental Change Series, Oxon, Routledge.

Fairhead, J., Leach, M. & Scoones, I. (2012) Green grabbing: A new appropriation of nature? *Journal of Peasant Studies*, 39 (2), 237–261.

FAO. (1997) *Regional Study on Wood Energy Today and Tomorrow in Asia*. Asia-Pacific Forestry Sector Outlook Study Working Paper Series. Food and Agriculture Organization of the United Nations (FAO), Rome and Bangkok. Working Paper No APFSOS/WP/34. Available from: http://wgbis.ces.iisc.ernet.in/energy/HC270799/RWEDP/acrobat/fd50.pdf

Goldemberg, J. & Lucon, O. (2009) *Energy, Environment and Development*. 2nd edition. Oxon, Earthscan, Routledge.

Haines, A. (2012) Health benefits of a low carbon economy. *Public Health*, 126 (1), 33–39.

Haines, A., McMichael, A.J., Smith, K.R., Roberts, I., Woodcock, J., Markandya, A. & Wilkinson, P. (2010) Public health benefits of strategies to reduce greenhouse-gas emissions: Overview and implications for policy makers. *The Lancet*, 374 (9707), 2104–2114.

Hao, H., Wang, H.W. & Ouyang, M.G. (2011) Comparison of policies on vehicle ownership and use between Beijing and Shanghai and their impacts on fuel consumption by passenger vehicles. *Energy Policy*, 39 (2), 1016–1021.

Hoekstra, A.Y. & Chapagain, A.K. (2007) Water footprints of nations: Water use by people as a function of their consumption pattern. *Water Resource Management*, 21, 35–48.

Hoekstra, A.Y. & Mekonnen, M.M. (2012) The water footprint of humanity. *Proceedings of the National Academy of Sciences*, 109, 3232–3237.

Kovats, S., Depledge, M., Haines, H., Fleming, L.E., Wilkinson, P., Shonkoff, S.B. & Scovronick, N. (2014). The health implications of fracking. *The Lancet*, 383(9919):757–758.

Kowalski, G. & Vilogorac, S. (2008) *Energy Security Risks and Risk Mitigation: An Overview*. UNECE Annual Report 2008. United Nations Economic Commission for Europe (UNECE). Available from: www.unece.org/fileadmin/DAM/oes/nutshell/2008/9_Energy_Security_Risks.pdf

Leggett, J. (2013) *The Energy of Nations: Risk Blindness and the Road to Renaissance*. Oxon, Earthscan, Routledge.

Majid Cooke, F., Saat, G., Gani, N. & Urban, F. (2014) Water-Food-Energy Security and its Social Implications. University of Malaysia Sabah, Kota Kinabalu.

Mielke, E., Daiz Anadon, L. & Narayanamurti, V. (2010) *Water Consumption of Energy Resource Extraction, Processing and Conversion*. Energy Technology Innovation Policy Research Group, Harvard Belfer Center, Harvard. Available from: http://belfercenter.ksg.harvard.edu/files/ETIP-DP-2010-15-final-4.pdf

Neville, K.P. & Dauvergne, P. (2012) Biofuels and the politics of mapmaking. *Political Geography*, 31 (5), 279–289.

Nwilo, P.C. & Badejo, O.T. (2008) Impacts of oil spills along the Nigerian coast. In: *Soil, Sediment & Water*. Available from: http://web.archive.org/web/20080430164524/www.aehsmag.com/issues/2001/october/impacts.htm

Odgaard, O. (2014) *China's Low Carbon Energy Policy: National Dilemmas and Global Perspectives*. Copenhagen, Copenhagen Business School.

Oxfam (2012) *The Hunger Grains: The Fight Is On. Time to Scrap EU Biofuel Mandates*. Oxfam. Available from: https://policy-practice.oxfam.org.uk/publications/the-hunger-grains-the-fight-is-on-time-to-scrap-eu-biofuel-mandates-242997

Parrish, D. (2013) Key issues and outlook. In: Zhu, T., Melamed, M.L., Parrish, D., Gauss, M., Gallardo Klenner, L., Lawrence, M., Konare, A. & Liousse, C. (Eds.) *WMO/IGAC Impacts of Megacities on Air Pollution and Climate*. GAW Report 205. pp. 285–299. Available from: www.wmo.int/pages/prog/arep/gaw/documents/GAW_205_DRAFT_13_SEPT.pdf

Rao, S., Pachauri, S., Dentener, F., Kinney, P., Klimont, Z., Riahi, K. & Schoepp, W. (2013) Better air for better health: Forging synergies in policies for energy access, climate change and air pollution. *Global Environmental Change*, 23 (5), 1122–1130.

Rathmann, R., Szklo, A. & Schaeffer, R. (2010) Land use competition for production of food and liquid biofuels: An analysis of the arguments in the current debate. *Renewable Energy*, 35 (1), 14–22.

Roper, L.D. (2019) *Fossil-Fuels Depletion*. L David Roper Interdisciplinary Studies. Available from: www.roperld.com/science/minerals/FossilFuelsDepletion.htm

Seinfeld, J.H. & Pandis, S.N. (2006) *Atmospheric Chemistry and Physics: From Air Pollution to Climate Change*. 2nd edition. Hoboken, NJ, Wiley.

Smithson, P., Addison, K. & Atkinson, K. (2008) *Fundamentals of the Physical Environment*. 4th edition. London, Routledge.

UK ERC. (2009) *The Global Oil Depletion Report*. UK Energy Research Centre (ERC). Available from: www.ukerc.ac.uk/publications/global-oil-depletion-an-assessment-of-the-evidence-for-a-near-term-peak-in-global-oil-production.html

UNFCCC. (1992) *United Nations Framework Convention on Climate Change*. United Nations, New York.

Urban, F. (2014) *Low Carbon Transitions for Developing Countries*. Oxon, Earthscan, Routledge.

US NRC. (2002) *Radioactive Waste: Production, Storage, Disposal. US Nuclear Regulatory Commission*. NUREG/BR-0216, Rev 2. Available from: www.nrc.gov/docs/ML1512/ML15127A029.pdf

Victor, N.M. & Victor, D.G. (2002) *Macro Patterns in the Use of Traditional Biomass Fuels. Program on Energy and Sustainable Development*. Stanford University, Stanford.

WHO. (2005) *Chernobyl: The True Scale of the Accident*. Geneva, World Health Organization (WHO). Joint News Release WHO/IAEA/UNDP. Available from: www.who.int/mediacentre/news/releases/2005/pr38/en/index.html

WHO. (2016) *Ambient (Outdoor) Air Quality and Health*. Geneva, World Health Organization (WHO). Factsheet. Available from: www.who.int/mediacentre/factsheets/fs313/en/

Wichelns, D. (2010a) Virtual Water and water footprints offer limited insight regarding important policy questions. *Water Resources Development*, 26 (4), 639–651.

Wichelns, D. (2010b) Virtual water: A helpful perspective, but not a sufficient policy criterion. *Water Resources Management*, 24 (10), 2203–2219.

12

THE ECONOMICS OF ENERGY SUPPLY AND UNIVERSAL ENERGY ACCESS

Energy supply: an economic perspective

The costs of providing fossil fuel energy

This section elaborates the costs of energy supply and investments for fossil fuels. Two issues need to be discussed here: the consumer price of fossil fuels and the actual cost of fossil fuels. Both depend on the investment costs of building fossil fuel power plants, as well as supporting infrastructure such as transmission and distribution lines; fuel costs to fire the power plants; operation and maintenance costs as well as insurance costs.

The consumer price of fossil fuels is what matters most to individuals, households, firms, etc. It is highly variable depending on each country, type of fossil fuel, type of power plant, quality, etc. Coal is usually the cheapest fossil fuel around the world, while natural gas and oil are more expensive. However, while some oil-based products are expensive in some countries, such as petrol in Germany or the UK, they are cheaper in other countries, such as in the USA. This is for a number of reasons, including access to fossil fuels (e.g. high availability in OPEC countries, hence cheaper prices), the profits energy firms and utilities make and the level of subsidisation. This leads us to the second issue: the actual cost of fossil fuels. This is strongly determined, among other factors, by fossil fuel subsidies and fuel import bills. We will discuss both issues here.

Fossil fuel subsidies

Although renewable energy has (wrongly) the reputation of being highly subsidised, this is, in fact, the case for fossil fuels. The IEA (2019a) calculates that in 2012 fossil fuel subsidies around the world amounted to US$544 billion, compared with

US$101 billion for renewable energy subsidies. Fossil fuels were therefore globally subsidised about five times more than renewable energy in 2012. In 2018, fossil fuel subsidies globally accounted for about US$400 billion (IEA, 2019). Data from the Organisation for Economic Co-operation and Development (OECD) suggest even higher figures, namely that about US$775 billion were spent on fossil fuel subsidies in 2012, while the International Monetary Fund (IMF) suggests that up to US$2 trillion fossil fuel subsidies could have been paid in recent years annually (OECD, 2013). There is clearly some uncertainty about the exact figure, but it is clear that fossil fuel subsidies are massive. More than 80% of the subsidies are spent in developing countries, whereas the remaining share is fossil fuel subsidies in industrialised countries and in global production. Most of the fossil fuel subsidies were for oil-based products, followed by natural gas and coal (OECD, 2013). Fossil fuels subsidies distort markets and energy prices. Highly subsidised fossil fuels are mainly cheaper than renewable energy because of their artificially lowered prices. Without these subsidies, fossil fuel prices would be much higher and would make renewable energy options more economically competitive. Iran topped the list of the highest spending on energy subsidies for 2018, with more than 20% of its GDP spent on fossil fuel subsidies, mostly for oil and natural gas (about US$26–27 billion each), but also for electricity (about US$17 billion) (IEA, 2019b). Saudi Arabia spent about 14% of its GDP on energy subsidies in 2018, namely about US$32 billion on fossil fuel subsidies (mainly oil) and more than US$13 billion on electricity subsidies. China spent about 13% of its GDP on energy subsidies in 2018, about US$18 billion on fossil fuel subsidies (mainly oil) and more than US$25 billion on electricity subsidies. Russia spent about 11% of its GDP on energy subsidies in 2018, mainly for natural gas (IEA, 2019b). The industrialised countries that spend the most of their gross domestic product (GDP) on fossil fuel subsidies are the USA, followed by Canada, Japan, Germany (mainly coal) and the UK. The USA spends more than 3% of its GDP on fossil fuels subsidies annually (IEA, 2019a).

Some Indian states spend as much as 50% of their annual state budget on electricity subsidies (Joseph, 2010). Lockwood (2013) reports that rather than benefitting poorer households, most of the benefits from fossil fuel subsidies are reaped by the middle class, with only 10% of the subsidies reaching the poor. Nevertheless, fossil fuel removal often results in widespread social opposition, frequently by poorer urban households. This is due to concerns that the money saved will not be spent on pro-poor activities, and the poor are disproportionately more affected by cuts than the wealthy. Fossil fuel subsidy removal therefore needs to be carefully arranged and alternatives put in place (Victor, 2009; Lockwood, 2013).

Yemen is a recent example of how fossil fuel subsidy removal needs to be done slowly and carefully to avoid severe impacts on the poor. The government in Yemen spent more than one fifth of its state budget on fossil fuel subsidies, in total US$3–3.5bn (£1.8–2.3bn) in 2013, thereby plunging into debt and taking away much needed funding from other important sectors, such as agriculture, food and water, education and health. The government decided in 2014 to drastically and quickly reduce its fossil fuel subsidies. It was reported that the price of diesel was increased by 95% (hence the price consumers paid earlier was in fact only 5% of the actual costs of diesel) and the price of gasoline increased by 65% (Guardian, 2014). Diesel and gasoline are used for transportation, but also for agriculture, such as for powering water pumps and tractors. This has led to severe impacts on the poor who are now faced with excessive fuel bills that they cannot pay. This is at the expense of other needs, thereby threatening food security, medical services and even schooling (e.g. poor people can no longer afford school books, pencils, etc.). The IMF called for government-led social protection measures that would help the poor to overcome the impacts of the fossil fuel subsidy removal; however, it was reported this has not happened in Yemen (Guardian, 2014). Yet, since then Yemen has plunged into a much more severe state of humanitarian crisis caused by violent conflict and the Saudi-led air strikes.

From an environmental perspective, fossil fuel subsidies need to be abolished to avoid dangerous climate change and to enable a low carbon transition. From a social perspective, however, a rapid and deep cut to fossil fuel subsidies is not necessarily recommended. Rather fossil fuel subsidies need to be removed in several incremental stages and over time, along with the introduction of social protection mechanisms for the poor to enable them to gain access to financial support and services. The money that would be made available from the phasing out of fossil fuels could be reinvested in low carbon energy options, such as renewable energy, and/or used for poverty reduction.

Costly fossil fuel imports

Another important issue is the import bill that countries face for accessing fossil fuels. In particular, countries that do not have domestic fossil fuel resources are struggling with these costs. Countries such as China, which has vast coal resources, or the OPEC countries, which have oil resources, are in a favourable position for keeping their energy bills low. Other countries, such as small island developing states, least developed countries and other poor countries with little or no fossil fuel resources find themselves in a tricky and costly situation as they need to import expensive fossil fuels. Figure 12.1 shows how less developed countries have had rapidly increasing oil import bills in the past due to increasing oil prices. Oil import bills alone made up around 5% of their GDP in 2011 (IEA, 2011).

The costs of providing low carbon energy

This section elaborates the costs of energy supply and investments for low carbon energy. There are two issues that need to be discussed here: the consumer price for low carbon energy and the actual cost for low carbon energy. Both depend on the investment costs of building low carbon power plants, as well as supporting infrastructure such as transmission and distribution lines; fuel costs to fire the power plants; operation and maintenance costs as well as insurance costs. Investment costs per MW are usually high for low carbon energy compared with fossil fuel power

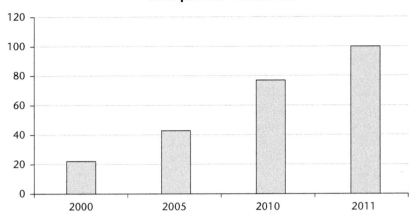

Oil Import Bill - Billion USD

FIGURE 12.1 Oil import bills for net importing, less developed countries

Source: IEA, 2011: 9

plants. However, while nuclear energy involves fuel costs, the fuel costs for renewable energy are zero as no fuel is used. Sun, wind and water are available for free at no fuel cost. This is a strong financial advantage over fossil fuels and nuclear energy, as in the long run the operating costs are much lower.

While the prices for nuclear energy remain high, the prices for renewable energy have decreased significantly since the 1980s. In 1985, solar silicone photovoltaic (PV) panels cost about US$7 per watt power (Wp) in OECD countries; in 2011, they cost less than US$2 per Wp. In China, prices are even lower and, in late 2011, solar silicone PV panels were below the US$1.00 per watt mark for the first time; by 2013 they were about US$0.50 per watt and have since declined further. The Nobel Prize–winning economics professor Paul Krugman suggested in 2011 that we were only a few years away from the point where electricity from solar energy will cost the same as electricity from fossil fuels, despite the high fossil fuel subsidies (Krugman, 2011). This phenomenon is called grid parity. In fact, this has happened in recent years (mostly around 2014) and today most types of renewable energy are cost competitive with fossil fuel prices, such as natural gas prices. Solar PV for example has seen a dramatic decline in prices and is now cost competitive with fossil fuel prices (Urban et al., 2018). Wind power too has become much more cost competitive in recent years (Partridge, 2018). In the past, it was assumed that technological learning and economies of scale were the main drivers of cost reductions for renewable energy technologies. Pillai (2015), however, suggests that for solar PV the driving factors are declines in the polysilicon price, changes in polysilicon usage, increasing conversion efficiency and industry investment. See Figure 12.2 for an overview of cost competitiveness of renewable energy technologies. The graph shows the rapid decline in the global levelised costs of electricity (LCOE) from

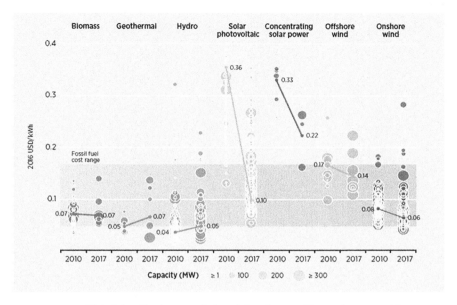

FIGURE 12.2   Global levelised cost of electricity from utility-scale renewable energy generation technologies, 2010–2017

Source: IRENA, 2018

solar PV and concentrating solar power (CSP), as well as onshore and offshore wind between 2010 and 2017. The figure also shows that today electricity generated from renewable energy is cost competitive with electricity generated from fossil fuels (hence achieving grid parity).

Particularly striking is the rapid price decline for solar PV since 2010.

Figure 12.3 provides an overview of the geographic differences in renewable energy costs across the world. The figure shows that the LCOE for renewable energy seemed to be the lowest in Asia and Europe and highest in Africa, but partly also in North America, Central America and the Caribbean and the Middle East. However, differences exist per technology; for example, the LCOE for hydropower is the most expensive in Europe but cheaper in all other world regions. This can be explained by the fact that much of Europe's hydropower generation capacity is located in the Nordic countries, such as Norway, Sweden, Finland, Iceland and other countries that have high electricity prices such as Switzerland. The cheapest hydropower generation can be found in Asia, mainly in China, according to the IRENA Renewable Energy Cost Database (2018). China hosts more than 30% of the global installed hydropower capacity. Chinese electricity prices are much cheaper than those in Europe.

While solar, wind and other renewable energies are also subsidised in some countries, such as through the feed-in tariff in many European countries, many other countries do not have subsidies for renewable energy. Feed-in tariffs are

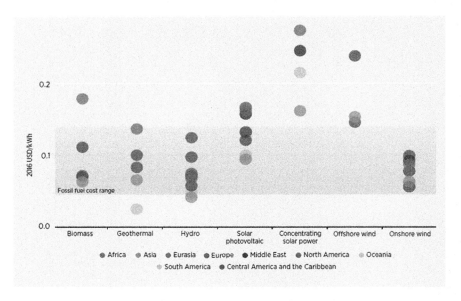

FIGURE 12.3 Regional weighted levelised cost of electricity (LCOE) by renewable energy generation technology for 2016/2017

Source: IRENA, 2018

a form of subsidy for renewable energy. They are a financial policy that aims to increase the share of renewable energy among the total energy mix. They offer long-term financial stability to renewable energy producers, usually in the form of the government paying a small amount of money per kWh electricity generated to the producers. The feed-in tariffs are usually specific for each type of renewable energy, so wind energy, for example, will receive a different tariff than solar PV. Several countries, such as Germany, Denmark, the UK, many other European countries, and even China and India, have feed-in tariffs.

For example, the Danish feed-in tariff for onshore wind energy was 3.35 EUR cent/kWh for the first 22,000 full-load hours plus 0.31 EUR cent/kWh for balancing costs totalling 3.66 EUR cent/kWh in 2014, and for offshore the fixed payment varies between different tenders, for example 6.95 EUR cent/kWh for the first 50,000 full-load hours. In Germany, the feed-in tariff for onshore wind energy in 2014 was 8.93 EUR cent/kWh for the first 5 years plus 0.48 EUR cent/kWh bonus totalling 9.41 EUR cent/kWh for the first 5 years, then 4.87 EUR cent/kWh for all following years. The feed-in tariff for offshore wind energy was in 2014 in Germany 15 EUR cent/kWh for the first 12 years, then 3.5 EUR cent/kWh or alternatively 19 EUR cent/kWh for the first 8 years (Lema et al., 2014). Note that these figures are regularly reviewed by government agencies and experts, hence they may be updated.

Marginal abatement cost (MAC) curves, also known as greenhouse gas (GHG) abatement cost curves, are used to calculate the annual carbon abatement potential for technological interventions and the costs of mitigation. These cost curves indicate that energy efficiency measures are the cheapest options for saving carbon dioxide emissions and thereby reducing the GHG emissions leading to climate change. As the MAC in Figure 12.4 indicates, other cost-effective low carbon energy options are geothermal energy, nuclear energy, solar PV and wind energy; whereas carbon capture and storage (CCS) is rather expensive (McKinsey & Company, 2010; Diesendorf, 2014). Figure 12.4 has been adjusted by McKinsey and Company for the global financial crisis.

To achieve a global low carbon economy to mitigate climate change, the IEA estimates that US$45 trillion of investments are required between 2010 and 2050 to reduce global carbon emissions by 50% – of which a significant amount will be needed for restructuring energy systems and moving to low carbon energy (IEA, 2010). A large share of this funding is expected to come from the private sector, although public finance – such as through overseas development assistance (ODA) – also plays a role as well as private–public partnerships.

The universal energy access case: an economic perspective

Different delivery models for universal energy access and their costs

In this section, we will discuss different ways of achieving universal energy access. We call these different ways delivery models. Delivery models for energy supply in this case refer to grid-connected or centralised versus off-grid or decentralised. Off-grid energy supply can be in the form of renewable energy technologies powering mini-grids, for example a micro-wind turbine or a micro-hydro plant that powers several homes in a village. Off-grid can also be in the form of stand-alone renewable energy technologies such as solar PV panels providing lighting and/ or electricity to individual homes or buildings. On-grid refers to grid-connected electricity access options. Different delivery models have different costs, which we will discuss here.

As we discussed in earlier units, the UN aims to achieve universal energy access for everyone everywhere by 2030 through their SE4All initiative. This includes access to electricity and access to clean cooking fuels. To remind you of the facts, it is estimated that about 1 billion people worldwide do not have access to electricity and about 2.7 billion people rely on traditional biomass – such as fuelwood and dung – for basic needs such as cooking and heating (IEA, 2019a). The largest number of people living without access to electricity in one single country is in India.

Investments in modern energy services are at the root of eradicating energy poverty. The IEA estimates that in 2009 about US$9 billion were invested globally in modern energy services, a figure that is expected to increase over the years. The IEA estimates that more than 30% of this funding came from multilateral organisations (e.g. the World Bank), 30% from domestic government funding, more than 20% from the private sector and about 15% from bilateral aid (IEA, 2011).

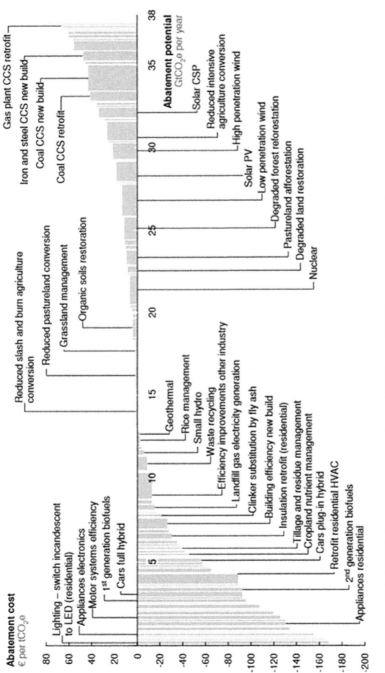

FIGURE 12.4 Global GHG abatement cost curve beyond BAU – until 2030

Source: McKinsey and Company (2010: 15)

TABLE 12.1 Different options for financing modern energy access

Technology solutions	Financing instruments	Types of financing sources
Electricity: on-grid, off-grid, mini-grid	Grants	Multilateral organisations
Cooking: LPG, biogas, advanced cook stoves	Equity	Bilateral official development assistance (ODA)
	Loans	Foreign direct investment (FDI)
	Insurance	Developing country governments
	Subsidies	Private sector initiatives
	Guarantees	

Source: Based on IEA (2011: 10)

To achieve universal energy access, the IEA estimates that US$1 trillion is needed, which is roughly US$48 billion per year, so about five times more than the investment levels for modern energy services in 2009. The IEA suggests that the majority of this funding has to come from multilateral and bilateral agencies, followed by domestic government funding and the private sector (IEA, 2011). Several financing instruments can be employed such as grants, private equity, loans, insurances, subsidies and guarantees.

Different options for financing modern energy access are shown in Table 12.1.

Electricity

We will now discuss in more detail what the economic implications of the universal energy access case are for delivering electricity.

The IEA estimates that about US$640 billion will be required to provide universal access to electricity worldwide, making up the majority of the total estimated US$1 trillion for the universal energy access case. The majority of the funding will be required in sub-Saharan Africa, followed by India (IEA, 2011). The IEA suggests that the average annual investment needed for electricity access under the UN's Sustainable Energy for All (SE4All) initiative between 2017 and 2030 will be over 25 billion USD in sub-Saharan Africa, roughly evenly split between costs for grid, mini-grid and off-grid. The IEA estimates that the annual costs for developing Asia will be much lower, at around 2–3 billion USD (IEA, 2017).

The universal energy access case can only be achieved by using different delivery models: *on-grid, mini-grid* and *off-grid*. On-grid plays a role where households are based in proximity to the grid and where it is economically and technically feasible to connect households to the grid. Mini-grid and off-grid particularly play a role in rural areas, especially in remote areas or mountainous areas where grid connections are not economically or technically feasible. According to the IEA (2011), most people could get access to electricity via on-grid or mini-grids, but a wide range of financing sources are needed, such as from the private sector, domestic governments and utilities, and multilateral and bilateral institutions, as well as microfinance schemes.

Mini-grids can either be from renewable energy, such as from hydro, wind or solar, or from fossil fuels, such as diesel, or they can be hybrid systems, for example a combination of renewables with fossil fuels for back-up capacity as needed.

Clean cooking fuels

We will now discuss in more detail what the economic implications of the universal energy access case are for delivering clean cooking fuels.

In terms of the annual investments in providing access to clean cooking facilities around the world between 2017 and 2030, the IEA (2017) suggest that the costs would be just below 25 billion USD in sub-Saharan Africa, just below 15 billion USD for India, around 7 billion USD for China, roughly 5 billion USD for Southeast Asia, about 7 billion USD for the rest of developing Asia and roughly 1–2 billion USD for the rest of the world. This shows the investment that is required to achieve universal energy access of clean cooking fuels according to the UN's SE4All initiative. A large proportion of the modern cooking fuels will be provided by investments in *biogas systems*, followed by *liquefied petroleum gas (LPG) stoves* and *advanced biomass cook stoves* (IEA, 2017).

The IEA (2011) suggests that while biogas systems are the most expensive way of proving modern cooking fuel, the largest number of people could likely get access to clean cooking fuels from advanced biomass cooking stoves and LPG. A wide range of financing sources is needed such as from the private sector, domestic governments, multilateral and bilateral institutions as well as microfinance schemes.

Universal energy access: fossil fuels versus renewable energy

Figure 12.2 shows electricity generation from different energy sources for universal energy access. The IEA (2011, 2017) reports that renewable energy is likely to account for about 90% of mini-grid and off-grid installations. Solar energy is likely to play a large role, followed by wind energy and biomass. For on-grid generation, fossil fuels are likely to account for more than half of the energy, mainly coal. Large-scale hydropower may play an important role in providing electricity for the grid too, while small-scale hydropower may play a role for off-grid electricity generation. The IEA describes hydropower as a renewable energy source that is cost competitive. While this is certainly the case for small hydropower, recent research highlights that the actual costs of large hydropower dams are often much higher than anticipated (Ansar et al., 2014). The research concludes that in the majority of countries large dams are too expensive to build and the planning and building process is too lengthy to result in positive investment returns (Ansar et al., 2014). Similar findings are made by Sovacool et al. (2014), who conclude that hydropower dams and nuclear power plants have the highest cost overruns of all types of power plants compared with their scheduled costs. This is coupled with the environmental and social impacts of large dams and their economic costs, such as compensation payments for relocation, costs of resettlement, loss of livelihoods, loss of biodiversity and damage to ecosystems, etc. While some of these impacts can be quantified rather easily, such as compensation payments, other impacts are difficult to quantify in economic terms, such as loss of biodiversity and flooding of ecosystems. Yet, it has to be remembered that some countries depend almost entirely on hydropower generation capacity for providing electricity access, such as Nepal, Mozambique and Norway. Most likely the alternatives would be fossil-fuel-powered plants, such as coal-based, if hydropower were not an option.

While these considerations relate to the economic implications of renewable energy, including large-scale hydropower, there are also several economic implications of using fossil fuels to provide universal energy access. It is often argued by critics of renewable energy that they are subsidised and not cost competitive. However, fossil fuels are heavily subsidised around the world, to a far greater extent than renewables. Fossil fuels are particularly subsidised by developing countries' governments. This puts severe financial strain on countries that are already poor and have a lack of financial resources to invest in other crucial sectors such as healthcare and education. Fossil fuels are globally subsidised about four to five times more than renewable energy (IEA, 2019a; IEA, 2019b; OECD, 2013). It is therefore clear that fossil fuels are not cheap fuels, and in contrast that a transition to low carbon energy for universal energy access, particularly renewables such as solar, wind and biomass could, in fact, be more cost effective.

Exercises

1   Choose a country of your choice. For this specific country, find out the levelised cost of electricity by renewable energy generation technology (see Figure 12.3 for a regional weighting as a proxy) and find out if the country has a feed-in tariff for renewable energy (or not) and how high the tariff is. What does this tell us about the competitiveness of renewable energy in your country of choice? Has grid parity been achieved yet?

2   Go to www.iea.org/weo/energysubsidies/. Choose one of the countries listed in the table on fossil fuel subsidies. How high are the fossil fuel subsidies per year for this specific country and as a percentage of GDP? What types of energy sources have been funded the most? What does this mean for the overall development of the country, particularly concerning economic, social and environmental implications?

References

Ansar, A., Flyvberg, B., Budzier, A. & Lunn, D. (2014) Should we build more large dams? The actual costs of hydropower megaproject development. *Energy Policy*, 69, 43–56.

Diesendorf, M. (2014) *Sustainable Energy Solutions for Climate Change*. Oxon, Earthscan, Routledge.

Guardian (26 August 2014) Yemen fuel subsidy cut drives poorest deeper into poverty. *Guardian Development Network, Guardian.* Available from: www.theguardian.com/ global-development/2014/aug/26/yemen-fuel-subsidy-cut-drives-poorest-poverty

IEA. (2010) *World Energy Outlook 2010. Energy Poverty: How to Make Modern Energy Access Universal?* Paris, International Energy Agency (IEA), OECD/IEA. Available from: www. worldenergyoutlook.org/media/weowebsite/2010/weo2010_poverty.pdf

IEA. (2011) *World Energy Outlook 2011. Energy for All. Financing Access for the Poor.* Paris, International Energy Agency (IEA), OECD/IEA. Available from: www.iea.org/media/ weowebsite/energydevelopment/weo2011_energy_for_all.pdf

IEA. (2017). *Energy Access Outlook.* Available from: https://www.iea.org/publications/ freepublications/publication/WEO2017SpecialReport_EnergyAccessOutlook.pdf

IEA. (2019a) *Statistics.* Paris, International Energy Agency (IEA), OECD/IEA. Available from: www.iea.org/statistics/

IEA. (2019b) *Fossil-Fuel Subsidies.* Available from: www.iea.org/weo/energysubsidies/

IRENA. (2018) *Renewable Power Generation Costs in 2017.* International Renewable Energy Agency, Abu Dhabi.

Joseph, K. (2010) The politics of power: Electricity reform in India. *Energy Policy*, 38, 503–511.

Krugman, P. (6 November 2011) Here comes the sun. The Opinion Pages, *The New York Times.* Available from: www.nytimes.com/2011/11/07/opinion/krugman-here-comes-solar-energy.html?hp&_r=0

Lema, R., Nordensvärd, J., Urban, F. & Lütkenhorst, W. (2014) *Innovation Paths in Wind Power: Insights from Denmark and Germany.* GDI/DIE, Bonn. Discussion Paper.

Lockwood, M. (2013) The political economy of low carbon development. In: Urban, F. & Nordensvärd, J. (Eds.) *Low Carbon Development.* Oxon, Routledge. pp. 25–37.

McKinsey and Company (2010) *Impact of the Financial Crisis on Carbon Economics.* New York, McKinsey.

OECD. (2013) *Inventory of Estimated Budgetary Support and Tax Expenditures for Fossil Fuels 2013.* Organisation for Economic Co-operation and Development (OECD), Paris. Available from: www.oecd-ilibrary.org/environment/inventory-of-estimated-budgetary-support-and-tax-expenditures-for-fossil-fuels-2013_9789264187610-en

Partridge, I. (2018) Cost comparisons for wind and thermal power generation. *Energy Policy*, 112 (1), 272–279.

Pillai, U. (2015) Drivers of cost reduction in solar photovoltaics. *Energy Economics*, 50 (7), 286–293.

Sovacool, B., Nugent, D. & Gilbert, A. (2014) Construction cost overruns and electricity infrastructure: An unavoidable risk? *The Electricity Journal*, 27 (4), 112–120.

Urban, F., Wang, Y. & Geall, S. (2018) Prospects, politics, and practices of solar energy innovation in China. *Journal of Environment and Development*, 27 (1), 74–98. Available from: https://doi.org/10.1177/1070496517749877

Victor, D. (2009) *The Politics of Fossil-Fuel Subsidies.* Series: Untold billions: fossil-fuel subsidies, their impacts and the path to reform. Global Subsidies Initiative (GSI), International Institute for Sustainable Development (IISD). Available from: www.iisd.org/gsi/sites/ default/files/politics_ffs.pdf

13

FINANCING LOW CARBON ENERGY TRANSITIONS

Carbon markets

Carbon markets aim to attach a price to carbon and create a trade of carbon emissions, usually measured in units of carbon dioxide or carbon dioxide equivalents (e.g. 1 tonnes of carbon dioxide or CO_2 equivalents) and traded, for example, in the form of certified emission reductions credits (CER). Carbon markets use carbon-offsetting mechanisms and set caps on carbon emissions to limit and reduce greenhouse gas (GHG) emissions either domestically, bilaterally or internationally. Bilateral means between two countries.

Carbon market mechanisms include the CDM, emissions trading, *joint implementation* and reducing emissions from deforestation and forest degradation (REDD). We will not discuss REDD here, as it is not directly related to energy issues (although there are concerns that monoculture plantations, such as palm oil plantations, may count as protectable forests under REDD, which creates an indirect link with contested biofuel developments).

The *Clean Development Mechanism (CDM)* enables industrialised countries to finance projects leading to emission reductions in developing countries. The CDM aims to contribute to sustainable development in developing countries and has to be 'additional', which means that investments should only be made for projects that create an additional GHG emission reduction, rather than funding already planned or proposed projects. The system works in a way that developing countries gain access to climate-friendly technology, such as renewable energy, while industrialised countries gain CERs to offset their emissions. In March 2019, nearly 8,000 projects had been registered with the UNFCCC and another 580 were pre-registered. More than 4,700 projects were dormant, since no contact was made with the UNFCCC secretariat since 2014 (UNEP, 2019a). The CDM has mainly contributed to wind energy, hydropower and biomass energy projects in developing countries, less in other forms of climate-friendly technology. It is estimated that the current CDM projects will account for over 1,028,800,000 CERs (tonnes of CO_2 equivalents) by 2020 (UNEP, 2019a). The large majority of CDM projects are located in China, India and other emerging economies, whereas poorer countries have failed to attract many CDM projects. Over 80% of all CDM projects are located in Asia, about half of them in China (UNEP, 2019a). However, as part of the second commitment period of the Kyoto Protocol, the CDM was reformed to enable poorer countries to benefit from the mechanism. Yet, only about 1.5% of all CDM projects are located in least developed countries and only about 0.5% in small island developing states. The countries that invested most often in CDM projects are the UK with nearly 2,500 CDM project investments, Switzerland with nearly 1,500 CDM project investments, the Netherlands with over 700 CDM project investments, and Japan and Sweden with each over 400 CDM project investments as of March 2019 (UNEP, 2019a).

A key criticism of the CDM is that it is an offsetting mechanism, meaning that a country or a firm invests in low carbon projects in poor countries, rather than reducing their own emissions. Urban and Nordensvärd (2013) argue that this might be considered a modern form of selling of indulgences. There is an ongoing debate over whether the CDM has contributed to technology transfer and knowledge transfer; however, the overwhelming conclusion is that this has been limited and that the main transfer has been of financial form, therefore generating investments in the recipient countries (Urban, 2014; Lederer, 2013).

The *Emissions Trading Scheme (ETS)* is a carbon market mechanism that caps (or limits) GHG emissions at a maximum level for specific industries and creates a trading system. Industries, firms and major institutional polluters get a maximum emission allowance, which limits, for example, their emissions from energy generation. Once this allowance is exceeded, emission credits must be purchased from others who have emitted less (Lederer, 2013; Urban & Nordensvärd, 2013). The world's first and largest ETS is the European Emissions Trading System (EU ETS). The EU ETS accounts for nearly half of all European CO_2 emissions. Each year about 6.5 Gt of carbon dioxide emissions are traded in the EU alone, which accounts for US$120 billion annually and 97% of global carbon trading (EC, 2005; Skjaerseth & Wettestad, 2008). Attaching a price to carbon means that polluting fossil fuels become increasingly expensive while renewable energy and other forms of low carbon energy become increasingly cost effective.

More globally, China has regional emission schemes in major cities including Beijing, Shanghai, Chongqing, Shenzhen, Tianjin, Guangdong Province and Hubei Province. California also has a regional ETS, as do other US states, as well as South Korea and Tokyo. New Zealand has an emission trading scheme, which has however been criticised as it has no emissions cap.

A major problem with the ETS is the problem of grandfathering. This was the case when the EU ETS was first set up. Grandfathering gets its name from the grandfather who cares for his children and grandchildren by means of inheritance. For the ETS, this means that governments are very protective of their national industries and therefore negotiate emissions caps that are far higher than needed. This, in turn, means that the price per ton of carbon is held artificially low, as there is an oversupply of carbon that can be traded. In early 2013, a ton of carbon dioxide was worth less than EUR 3 in the EU ETS, while experts calculate a tonne of carbon dioxide needs to be traded at EUR 20–30 to achieve a reduction of emissions, a level that was attained between 2006 and 2008. However, with each new phase of the EU ETS, excessive emission allowances were an issue, leading to a decline in the price of tradable carbon emissions due to an oversupply. This problem could be rectified by the back-loading approach, which aims to take up to 900 million carbon allowances out of the EU ETS. However, senior politicians are hesitant and still manage to negotiate exceptions for their industries and block international decisions (see Business Green Plus, 2013; Stonington, 2013). Nevertheless, as part of the second phase of the Kyoto Protocol for the period 2013 to 2020, a tighter EU emissions cap was set and there was a move from emission allowances to emission auctioning.

Joint implementation (JI) is another carbon-offsetting scheme that allows industrialised countries (so-called Annex I countries under the Kyoto Protocol) to invest in CER projects in other industrialised countries as an alternative to reducing emissions domestically. It is a mechanism for industrialised countries only. For example, if one Annex I country has reduced their emissions since 1990 and another Annex I country has exceeded their emissions, the latter country can offset their emissions in the former by investing in emission reduction projects in that country. Most JI projects are happening in the sectors of non-carbon dioxide emissions avoidance (e.g. fugitive emissions, landfill, methane avoidance, nitrous oxide avoidance), energy efficiency (particularly industrial energy efficiency) and renewable energy (particularly wind and biomass). By March 2019, about 790 JI projects and nearly 90 Programmes of Action (PoA) were recorded with a total of more than 863,500,000 emission reduction units (ERU) (ton of CO_2 equivalents) (UNEP, 2019b). Russia and Ukraine have the highest number of JI projects, hence most of the investment for JI are received in these two countries, followed by various other Eastern European countries. This can be explained by the sharp decline in emissions that the former Eastern Bloc countries experienced after the collapse of the USSR. This was driven by a sharp decline in the industrial sectors of the former Eastern Bloc countries during the 1990s and the modernisation of power plants and industrial plants in subsequent years. The largest buyers of JI ERUs are the Netherlands that invested in 200 JI projects, Switzerland that invested in nearly 140 JI projects, followed by the UK (with more than 70 JI project investments), Latvia (about 45 JI project investments) and Germany (nearly 40 JI project investments) as of March 2019 (UNEP, 2019b).

While many scholars and analysts agree that the market-based mechanisms of the ETS, CDM and JI are powerful tools to reduce emissions, others are sceptical. Bachram (2004) even refers to 'climate fraud' and 'carbon colonialism' when discussing carbon markets, because (a) emissions are not being reduced at their sources but rather offset such as through the CDM, JI and ETS and (b) low carbon investments are being made by richer countries in poorer countries, e.g. in the global South.

Bilateral and multilateral funding

The most common form of funding for energy access and low carbon energy transitions comes from national and local governments. Virtually every country in the world has some funds and schemes for investing in energy. This can take the form of rural electrification, as is the case in many Asian and African countries, state-owned utilities, which generate electricity in countries such as in Latin America, Asia, Africa and Europe, investments in grid infrastructure, fossil fuel subsidies, feed-in tariffs, investments in R&D, piloting demonstration projects for new energy technology, etc.

Another source of investment comes from bilateral (between two countries) and multilateral (between several countries) funding. Bilateral funding for energy access and low carbon energy transitions comes from so-called donor governments, which donate money to poorer countries. Traditionally, donor organisations (or aid organisations) were based in industrialised countries, in Europe, the USA, Australia and Japan. They would operate through donor organisations such as the Department for International Development (DFID), Deutsche Gesellschaft für Internationale Zusammenarbeit GmbH (GIZ), Norwegian Agency for Development Cooperation (Norad), Swedish International Development Cooperation Agency (SIDA), Danish Development Agency Danida, United States Agency for International Development (USAID), Australian aid agency AusAid, the Japan International Cooperation Agency (JICA) and others. Increasingly, emerging economies such as China, India and Brazil are also donors. They tend to operate in somewhat different ways to 'traditional' donors and their approach to aid tends to be more complex and decentralised, as they do not have classical donor organisations.

Key multilateral funding for low carbon energy transitions comes from the UNFCCC, for example the Green Climate Fund, as well as from the World Bank and the Global Environmental Facility (GEF). We will discuss these funding sources below.

The Green Climate Fund has its roots in discussions at the Conference of the Parties 15 (COP 15) in Copenhagen in 2009, and was formalised as part of the Cancun Agreements at COP 16 in 2010, then launched at COP 17 in Durban in 2011. The Green Climate Fund includes US$30 billion 'fast track' finance from industrialised countries to poor countries for the period 2010–2012, and medium-term finance of US$100 billion annually by 2020. The Green Climate Fund is aimed at helping the most vulnerable countries (least developed countries, LDCs, African countries and small island developing states, SIDS) to adapt to climate change, provide access to climate mitigation technology, such as renewable energy technology, and fund REDD activities. The fund also finances technology transfer, technology development and capacity building. The funding should be 'new and additional' to existing official development assistance (ODA) to avoid funding being taken from other important development sectors, such as healthcare and education. However, the notion of 'new and additional' has been contested, as several country governments mentioned they would redirect funding for the Green Climate Fund from other ODA budgets (Urban & Nordensvärd, 2013). The Green Climate Fund has been slow to be operationalised and, at COP 19 in Warsaw, it was agreed that the fund should start disbursements from 2015. Concerns have been raised that the pledges made by industrialised countries are below the amount the Green Climate Fund was supposed to raise, a problem which has also been observed with the Adaptation Fund and the Least Developed Country Fund. By 2011, the actual amount that has been pledged to the Green Climate Fund by donor countries was about US$12 billion, which is far below the pledged US$30 billion (Polycarp, 2011). The Paris Agreement of 2015 includes climate finance of US$100 billion annually which is supposed to be paid by developed countries for climate change adaptation and mitigation in developing countries during the time 2020 to 2030.

Other funding for energy access and low carbon energy transitions includes funding from donor countries that is being administered by the multilateral organisation World Bank. These include the Climate Investment Fund (CIF) which includes within its four programmes two focussed on energy access and low carbon technology: the Clean Technology Fund (CTF), and the Scaling-Up Renewable Energy Programme (SREP) in low-income countries and two other schemes that are more focussed on climate change issues. Urban and Nordensvärd (2013) report:

> the external funding – particularly the World Bank funding – has not been particularly well received by the G77 (the group of 77 powerful developing countries) and it is seen as potentially undermining the legitimacy of the UNFCCC in tackling climate change. The G77 has argued over recent years that all climate funding should be under the control of the UNFCCC, hence the UNFCCC's Green Climate Fund is of outmost importance.
>
> *Urban and Nordensvärd (2013: 86)*

Of particular interest to the energy debate are the CTF and the Strategic Climate Fund (SCF), which are part of the World Bank's CIF. The CIF has leveraged finance of over US$6 billion and is funded by Organisation for Economic Co-operation and Development (OECD) countries (World Bank, 2012). The Clean Technology Fund provides funding to finance the demonstration, deployment and transfer of low carbon technologies for mitigation in the power sector, the transport sector and energy efficiency. The funding is made available through concessional financing instruments, such as grants and concessional loans, and risk mitigation instruments (World Bank, 2008a). The Strategic Climate Fund provides funding for pilot projects and technology transfer in the fields of renewable energy, forestry and climate resilience (World Bank, 2008b, 2008c). Funding eligibility under World Bank rules is for countries that qualify for ODA and have a lending programme and/or policy dialogue with the World Bank or other multilateral development banks (Urban & Nordensvärd, 2013).

The Scaling-Up Renewable Energy Program for LDCs is another funding programme that is of interest to the energy access and low carbon energy transitions debate. The budget of SREP is around US$250 million, coming from various OECD donors. Urban and Nordensvärd (2013: 87) report that the greatest barriers to renewable energy development in poor countries are 'a lack of access to capital, the need to engage the private sector to increase investments in renewable energy, a lack of affordability of current technologies and weak enabling environments'. The SREP aims to overcome these barriers by helping to scale up the deployment of renewable energy, encouraging energy access, particularly access to electricity, and mainstream renewable energy in national energy provision.

> The fund focuses both on providing access to electricity for the poor and promoting income generating activities. Small-scale renewable energy technology is being promoted such as biomass energy technology, solar, hydro, wind and geothermal energy technology, with an average capacity of below 10 MW. Both off-grid and grid-connected applications are being promoted for electricity generation to households and industries. Thermal renewable energy for industry, agriculture, commerce and households are also promoted. Financing is achieved through aid, grants, loans and credit enhancement (World Bank 2009).
>
> *Urban and Nordensvärd (2013: 87)*

Public, private and hybrid funding for low carbon energy transitions

The sections above mainly discussed public funding from states, at either the bilateral or the multilateral level. In addition to the funding streams identified above, public funding can also take the form of investments or foreign direct investments (FDI, if the investment is in another country), ODA (as a form of aid) or as a trade deal.

In addition to state and bilateral/multilateral funding options for low carbon energy, there is a wide range of private funding options. The International Energy Agency (IEA) reports that for halving emissions by 2050, about US$45 trillion will be required for investing in low carbon technologies and a large share will have to come from the private sector (IEA, 2010). Private firms, such as energy technology firms like solar and wind firms, play an important role in financing R&D for low energy carbon technology innovation, investing in low carbon technology and their deployment.

There is also a hybrid form of funding, which involves cooperation between public and private actors. This is known as a PPP, a public–private partnership. One major advantage of PPPs is that private investments are made available (often beyond the scale of state funding), yet they often serve the public good, as the state is part of the funding deal and has the broader goal to act in the public's interest. This form of funding is also supposed to increase accountability, as private investors can be made accountable by state involvement. This is increasingly a form of funding partnership favoured by many funding agencies, project developers and government agencies. Yet, as always, there are exceptions to the rule and it depends on each specific case whether these theoretical rules apply in reality.

Newell (2011: 94) differentiates between different forms of governance of energy finance, namely: '(i) the public governance of public finance; (ii) the public governance of private finance; and (iii) the private governance of private finance'.

The cost of climate change and mitigation

The Stern Review on the economics of climate change

The Stern Review: key findings

The famous Stern Review on the economics of climate change, commissioned by the Treasury of the UK government, discusses the economic impacts of climate change and the costs of stabilising GHG emissions at levels that avoid dangerous climate change (Stern, 2006, 2007). The Stern Review indicates that to avoid dangerous climate change, global emissions need to be stabilised at 550 ppm CO_2e. However, leading climate scientists estimate that for a 50% chance of achieving the 2 °C target to avoid dangerous climate change, a global atmospheric carbon dioxide equivalent concentration of 400–450 ppm needs to be achieved by 2100 (Richardson et al., 2009; Pye et al., 2010). Stabilising CO_2e concentration at 550 ppm is more likely to result in a 3 °C warming, rather than a 2 °C warming, according to recent scientific studies. Despite these pitfalls, the Stern Review reports that global emissions should peak roughly between 2015 and 2030, and by 2050 global emissions need to have declined by at least 25% compared with current levels (Stern, 2006; Urban & Nordensvärd, 2013). When we compare this to actual emissions, there is no indication that global emissions are declining yet (IEA, 2019).

According to the Stern Review, unmitigated climate change could cost between 5% and 20% of global gross domestic product (GDP) annually due to the costs of the damage from sea-level rise, extreme weather events, rising temperatures, etc. (Stern, 2006). The Stern Review suggests that stabilising global emissions at 550 ppm CO_2 equivalents could cost about 1% of global GDP annually by 2050. Stern describes this as a cost that is 'significant but manageable' (Stern, 2006: xii). The Stern Review hails market mechanisms, such as emission trading schemes (ETS), as the cheapest option for reducing emissions (Stern, 2007). According to the Stern Review, the damage costs of every tonne of carbon dioxide are approximately US$85, whereas the cost for reducing emissions would be less than US$25 per tonne (Stern, 2006; Urban & Nordensvärd, 2013).

The Stern Review therefore sees opportunities in climate change mitigation rather than challenges. The report encourages the use of low carbon energy and low carbon technology as well as the creation of markets for low carbon products and services. It estimates that by 2050 a global low carbon technology market could be worth about US$500 billion (Stern, 2006; Urban & Nordensvärd, 2013).

Limitations of the Stern Review and criticism

While the Stern Review was lauded by the wider scientific community as an important work that highlighted the cost of inaction towards climate change and encouraged a public debate, the report was criticised by others.

First, there is some scientific uncertainty concerning how exactly climatic change will unfold in the future, which largely depends on national, regional and global future emissions trajectories, and hence there is some uncertainty about the exact costs of climate change. Rather than stating specific figures, the report therefore indicates scientific uncertainty in this debate by reporting a range of figures with low and high estimates. The report also makes clear that the figures are based on estimates and describe potential events, rather than being predictions. The methodologies used by the Stern Review include economic models, such as macroeconomic nature and integrated assessment models, as well as disaggregated approaches (Urban & Nordensvärd, 2013).

Regarding key criticism of the Stern Review, Urban and Nordensvärd (2013) report:

> the figures for the financial damage of climate change could be estimated too high, while the figures for the cost of mitigation could be estimated too low. Other well-known studies suggest that the financial damage of climate change is in the range of +1 to −5% of global GDP each year (compared to −5% to −20% of global GDP advocated by the Stern Review) and the cost of mitigation is estimated in the range of 0 to −7.5% of global GDP (compared to +1 to nearly −4% of global GDP suggested by the Stern Review) (Dasgupta, 2006; Tol, 2006; Tol & Yohe, 2006; Dietz et al., 2007; Mendelsohn, 2007; Weitzman, 2007; Helm & Hepburn, 2009; OECD, 2009). In addition, the Review's methodology was criticised as it works with existing predictions of climate impacts and existing cost estimates. The study does not calculate new figures from scratch, but uses existing figures from earlier studies.
>
> *Urban and Nordensvärd (2013: 88)*

The criticism regarding the financial damage from climate change impacts relates partly to the issue of a low discount rate. The discount rate suggests that a given sum of money today is generally considered more valuable than the same amount of money in the future. This is based on the facts that (a) future generations are assumed to be richer than today's generation due to economic growth and (b) people value money more if it is given to them today rather than in the future. This has the following implications: (a) A high discount rate therefore assumes that economic growth will increase substantially in the future. This assumption might have been correct before the financial crisis in 2008; however, due to the continuing recession and lower economic growth in many parts of the world, a lower discount rate might be more sensible as it implies lower economic growth. (b) The Stern Review does not value the welfare of current generations differently than the welfare of future generations, hence implying an element of intergenerational justice in relation to climate change. Both assumptions were criticised by other studies (see e.g. Weitzman, 2007; Giddens, 2009; Helm & Hepburn, 2009; Urban & Nordensvärd, 2013: 88–89). Another mentioned criticism is related to the fact that climate change adaptation strategies were not considered in the report.

Despite this criticism, there is strong support from climate scientists and climate economists for the findings and the message of the Stern Review.

> The Stern Review remains the most important and most cited report on the economics of climate change and it is commonly accepted nowadays that actions that delay mitigation will not only risk dangerous climate change, but also entail costs that will be higher in the long run than the costs for mitigating climate change.
>
> *Urban and Nordensvärd (2013: 89)*

The new climate economics report

In late 2014, a new, updated report on climate economics was published by Stern and his colleagues from the Global Commission on the Economy and Climate, entitled the New Climate Economy Report. The report is supported by the World Bank, the International Monetary Fund (IMF), the United Nations and the OECD and new annual reports are being published each year.

The New Climate Economy Report suggests that economic growth and reducing carbon emissions for mitigating climate change are compatible with each other. Stern and colleagues argue that a low carbon development model can in fact generate economic growth (Global Commission on the Economy and Climate, 2014). Other scholars are, however, more sceptical of this neoliberal approach and question whether economic growth and carbon emission reductions can be achieved at the same time.

The report argues that this is only possible if an adequate price is attached to carbon. This also requires switching to low carbon energy technology and energy-efficient technology at a large scale and on a global level. The report also suggests other measures such as more sustainable, low carbon city planning, stopping deforestation and promoting afforestation, as well as phasing out fossil fuel subsidies. The report suggests that the global economy needs to be transformed within the next 15 years to mitigate climate change, hence by the 2030s. A low carbon transition (or 'transformation', as the report puts it) can also lead to co-benefits such as reducing air pollution and helping to reduce poverty (Global Commission on the Economy and Climate, 2014).

While the report uses a neoliberal perspective and follows to some extent the idea of a 'technical fix', it provides much needed advice on how to transform high carbon economies at a global level to mitigate climate change.

Each year, the Global Commission on the Economy and Climate published a new report under the New Climate Economy Report label. The 2016 version dealt with sustainable infrastructure, for example.

More insights about opportunities for funding low carbon energy transitions can also be found in the International Finance Corporation's (IFC) 2017 report on climate investment opportunities. In common with the Stern Report and the New Climate Economy Reports, the IFC (2017) argues that climate change is not only a liability, but also opens up investment opportunities. The IFC Report links the national commitments made by the signatories of the Paris Climate Agreement to investment opportunities. It finds that 21 emerging market economies are planning to invest a total of US$23 trillion in climate-relevant markets for the achievement of their GHG emission reduction targets by 2030. This covers a range of sectors, such as energy generation (both grid-connected and off-grid), agriculture, transport, buildings, water infrastructure and waste management (IFC, 2017).

The link to energy and development

You may be wondering what the Stern Review on the economics of climate change and the New Climate Economy Report have to do with energy and development issues. We have discussed throughout this book that GHG emissions from energy-related activities make up the largest share of emissions, surpassing other GHG sources, such as land-use change, transport or industry. Understanding the basics of the economics of climate change is therefore important for understanding how energy supply and demand can contribute to mitigating climate change while taking economic implications into consideration. The Stern Review, the New Climate Economy Report and other leading studies conclude that mitigating climate change early is far more cost effective than continuing business-as-usual or mitigating at a later stage. For energy issues, this means that in the long run, it will be more cost effective to invest in low carbon energy technology and to increase the share of low carbon energy rather than continuing a heavy reliance on fossil fuel energy and fossil fuel technology for decades. This involves key energy issues, such as promoting investments and research and development (R&D) in low carbon energy technology and infrastructure (e.g. charging stations for electric vehicles), shifting form fossil fuels to renewables and reducing fossil fuel subsidies.

From a development perspective, this means the following: Poorer countries are disproportionately affected by climate change due to their high exposure and their limited financial, infrastructural and organisational capacities to adapt to climatic impacts. At the same time, poorer countries have less access to modern technology for climate change mitigation, such as wind turbines, solar panels or geothermal power stations. This is not only a question of access to technology, but also a question of access to investments and financing. As the Stern Review on climate change economics implies, financial assistance (such as ODA) could help poorer countries mitigate climate change and provide access to modern low carbon energy technology. Another domestic financing scheme is available through reducing fossil fuel subsidies and using (parts of these) as renewable energy/low carbon energy subsidies or for financing climate change adaptation. Finally, the findings of the Stern Review and the New Climate Economy Report can be related to the energy–poverty–climate nexus. Low carbon energy technology is not only a way of reducing emissions and thereby mitigating climate change, but for poor countries it can be an opportunity to provide modern energy access to those living in energy poverty. As the New Climate Economy Report points out, there are co-benefits to low carbon transitions, such as reducing air pollution and contributing to poverty alleviation.

Exercises

1   Review the different financing options for low carbon energy transitions. Make a list of their similarities and differences.
2   Go to the UNEP DTU CDM Pipeline Analysis and Database at www. cdmpipeline.org/. Look at the two datasets on CDMs. Choose a country and look into the dataset in detail for this specific country. What types of CDM project have been funded? Which countries/firms invested in these projects? How much CO_2 has been mitigated through these projects?
3a Go to the World Bank open dataset at https://data.worldbank.org/. Choose the figures for global GDP. The Stern Review suggests that unmitigated climate change could cost between 5% and 20% of global GDP annually due to the costs of the damage from sea-level rise, extreme weather events, rising temperatures, etc. (Stern, 2006). The Stern Review suggests that stabilising global emissions at 550 ppm CO_2 equivalents could cost about 1% of global GDP annually by 2050. Calculate the impacts of costs from climate damage vs the costs for climate change mitigation for global GDP.
3b Other well-known studies suggest that the financial damage of climate change is in the range of +1 to −5% of global GDP each year (compared to −5% to −20% of global GDP advocated by the Stern Review) and the cost of mitigation is estimated in the range of 0 to −7.5% of global GDP (compared to +1 to nearly −4% of global GDP suggested by the Stern Review) (Dasgupta, 2006; Tol, 2006; Tol & Yohe, 2006; Dietz et al., 2007; Mendelsohn, 2007; Weitzman, 2007; Helm & Hepburn, 2009; OECD, 2009). Calculate these figures for global GDP and make a judgment of the differences between these figures by other authors and the Stern Review figures.
3c Assume you are a policymaker responsible for climate and energy policy. Based on your calculations in 3a and 3b, what would you suggest to parliament with regards to climate (in)action?

References

Bachram, H. (2004) Climate fraud and carbon colonialism: The new trade in greenhouse gases. *Capitalism Nature Socialism*, 15 (4), 5–20.

Business Green Plus (2013) *Why Angela Merkel Holds the Key to Saving the World's Carbon Market*. Available from: www.businessgreen.com/bg/analysis/2320796/why-angela-merkel-holds-the-key-to-saving-the-worlds-carbon-market

Dasgupta, P. (2006) *Comments on the Stern Review's Economics of Climate Change*. University of Cambridge (UK), 12 December 2006. Published as: Commentary: The Stern Review's economics of climate change. *National Institute Economic Review*, 199, 4–7.

Dietz, S., Hope, C. & Patmore, N. (2007) Some economics of 'dangerous' climate change: Reflections on the Stern Review. *Global Environmental Change*, 17 (3–4), 311–325.

EC. (2005) *MEMO/05/84 Questions and Answers on Emissions Trading and National Allocation Plans*. Brussels, European Commission (EC).

Giddens, A. (2009) *The Politics of Climate Change*. New York, Wiley.

Helm, D. & Hepburn, C. (2009) *The Economics and Politics of Climate Change*. Oxford, Oxford University Press.

IEA. (2010) *World Energy Outlook 2010. Energy Poverty: How to Make Modern Energy Access Universal?* Paris, International Energy Agency (IEA), OECD/IEA. Available from: www.worldenergyoutlook.org/media/weowebsite/2010/weo2010_poverty.pdf

IEA. (2019) *Statistics*. Available from: https://www.iea.org/

IFC. (2017) *Creating Markets for Climate Business – An IFC Climate Investment Opportunities Report*. International Finance Corporation (IFC). Available from: www.ifc.org/wps/wcm/connect/974eedcb-f3d9-4806-b32e-73720e6f4ca7/IFC-Climate_Investment_Opportunity_Creating_Markets.pdf?MOD=AJPERES

Lederer, M. (2013) The future of carbon markets: Carbon trading, the Clean Development Mechanism and beyond. In: Urban, F. & Nordensvärd, J. (Eds.) *Low Carbon Development: Key Issues*. Oxon, Earthscan, Routledge. pp. 94–106.

Mendelsohn, R.O. (2007) A critique of the Stern Report. *Regulation*, 29 (4), 42–46.

Newell, P. (2011) The governance of energy finance: The public, the private and the hybrid. *Global Policy*, 2 (s1), 94–105.

OECD. (2009) *The Economics of Climate Change Mitigation. Policies and Options for Global Action Beyond 2012*. Organisation for Economic Co-operation and Development (OECD), Paris. Available from: www.oecd.org/env/cc/theeconomicsofclimatechangemitigation policiesandoptionsforglobalactionbeyond2012.htm

Polycarp, C. (2011) *Have Countries Delivered Fast Start Climate Finance?* Washington DC, World Resources Institute (WRI). Available from: www.wri.org/blog/2011/05/have-countries-delivered-fast-start-climate-finance

Pye, S., Watkiss, P., Savage, M. & Blyth, W. (2010) *The Economics of Low Carbon, Climate Resilient Patterns of Growth in Developing Countries: A Review of the Evidence*. Stockholm Environment Institute (SEI) Report to UK Department for International Development (DFID). Available from: http://sei-international.org/mediamanager/documents/Publications/Climate/economics_low_carbon_growth_report.pdf

Richardson, K., Steffen, W., Schellnhuber, H.J., Alcamo, J., Barker, T., Kammen, D.M., Leemans, R., Liverman, D., Munasinghe, M., Osman-Elasha, B., Stern, N. & Wæver, O. (2009) *Synthesis Report. Climate Change. Global Risks, Challenges and Decisions*. 10–12 March 2009, Copenhagen. University of Copenhagen. Available from: www.pik-potsdam.de/news/press-releases/files/synthesis-report-web.pdf

Skjaerseth, J.B. & Wettestad, J. (2008) *EU Emissions Trading*. Initiation, Decision-Making and Implementation. Aldershot, Ashgate.

Stern, N. (2006) Executive summary. In: *Stern Review: The Economics of Climate Change*. Cambridge, Cambridge University Press. Available from: http://webarchive.nationalarchives.gov.uk/20130129110402/www.hm-treasury.gov.uk/d/Executive_Summary.pdf

Stern, N. (2007) *The Economics of Climate Change. The Stern Review*. Cambridge, Cambridge University Press.

Stonington, J. (25 January 2013) *Wake-Up Call: A Disastrous Week for Carbon Trading*. Spiegel Online International. Available from: www.spiegel.de/international/europe/drop-in-carbon-price-underscores-disastrous-week-for-carbon-trading-a-879769.html

The Global Commission on the Economy and Climate (2014) The executive summary. In: *Better Growth, Better Climate: The New Climate Economy Report*. Available from: http://new climateeconomy.report/wp-content/uploads/2014/08/NCE_ExecutiveSummary.pdf

Tol, R. (2006) The Stern Review of the economics of climate change: A comment. *Energy and Environment*, 17 (6), 977–981.

Tol, R. & Yohe, G. (2006) A review of the Stern Review. *World Economics*, 7 (4), 233–250. Available from: http://qed.econ.queensu.ca/pub/faculty/garvie/econ443/debate/tol% 20and%20yohe%20review%20of%20stern%20review.pdf

UNEP. (2019a) *UNEP DTU CDM Pipeline Analysis and Database*. Available from: www. cdmpipeline.org/ UNEP DTU Partnership (Formerly UNEP Risoe), United Nations Environment Programme (UNEP), Denmark.

UNEP. (2019b) *UNEP DTU JI Pipeline Analysis and Database*. Available from: www. cdmpipeline.org/ji-projects.htm UNEP DTU Partnership (Formerly UNEP Risoe), United Nations Environment Programme (UNEP), Denmark.

Urban, F. (2014) *Low Carbon Transitions for Developing Countries*. Oxon, Earthscan, Routledge.

Urban, F. & Nordensvärd, J. (Eds.) (2013) *Low Carbon Development: Key Issues*. Oxon, Earthscan, Routledge.

Weitzman, M.L. (2007) A review of the Stern Review on the economics of climate change. *Journal of Economic Literature*, 45 (3), 703–724.

World Bank (2008a) *The Clean Technology Fund*. Washington DC, The World Bank.

World Bank (2008b) *Strategic Climate Fund (SCF)*. Washington DC, The World Bank. Available from: http://siteresources.worldbank.org/INTCC/Resources/Strategic_Climate_ Fund_final.pdf

World Bank (2008c) *Governance Framework for the Strategic Climate Fund*. Washington DC, The World Bank.

World Bank (2009) *Strategic Climate Fund Program SCF for Scaling-up Renewable Energy in Low Income Countries (SREP)*. Climate Investment Funds. Washington DC, The World Bank. Preliminary Design Document. CIF/SREPWG/2. Available from: www. climateinvestmentfunds.org/cif/sites/climateinvestmentfunds.org/files/SREPWG_ Preliminary_Document.pdf

World Bank (2012) *Strategic Framework for Development and Climate Change*. The World Bank, Washington DC. Available from: https://openknowledge.worldbank.org/bitstream/han dle/10986/13222/762400V10ESW0P00Box374367B00PUBLIC0.pdf?sequence=1&is Allowed=y

14

TECHNOLOGY FOR ENERGY AND DEVELOPMENT

Fossil fuels

Technological advances: fossil fuels

State-of-the-art fossil fuel technology

This book discusses fossil fuel technology first, before discussing low carbon energy technology in the next chapter. This is, however, with the clear statement that, in the long-term fossil fuels, need to be phased out and replaced by renewable energy for mitigating climate change. This is not a futuristic fantasy, but in line with many countries' policy plans. For example, the European Union aims to achieve carbon neutrality by 2050 (EC, 2018). Other countries have even more ambitious plans, such as Sweden that aims to have 100% of its electricity from renewables by 2040 (IEA, 2019a).

However, today more than 80% of the world's primary energy supply still comes from fossil fuel resources such as oil, coal and natural gas. Oil consumption for the transport sector plays a major role for the global primary energy supply. About 60% of the global electricity supply comes from fossil fuels (IEA, 2019b).

Coal, such as brown coal or lignite, is mainly used for heating, cooking and electricity generation. The energy efficiency of coal is lower than that of natural gas and oil. Coal is largely used for base-load electricity production. It has a higher emission factor than oil and natural gas, is more polluting and contributes substantially to global warming and air pollution. Many poorer people are still using coal as the preferred fuel for cooking and heating, even in upper-middle-income countries and emerging economies such as China. There are conventional coal-fired power plants and more efficient, ultra-super-critical coal power plants such as the integrated gasification combined cycle (IGCC) power plants. Their efficiencies are much higher than the average efficiency for a conventional coal-fired power plant. We will talk about IGCCs later.

Oil and oil-based products such as petrol, diesel, kerosene and liquefied petroleum gas (LPG) are mainly used for transport. Oil is sometimes used for heating and for generating electricity, although its energy efficiency is lower than that of natural gas. Oil is mainly used for base-load electricity production, although it can be used for peak-load production too. It has a higher emission factor than natural gas, but a lower emission factor than coal, which means that it contributes more to global warming than natural gas does, but less than coal. Conventional oil-fired power plants are more efficient than coal-fired power plants, but less efficient and less flexible than natural-gas-fired power plants. The majority of today's electricity comes from coal or natural gas, whereas oil is mainly used for transport purposes, and oil-based products, such as kerosene or diesel, are used for running private generators rather than large-scale electricity generation from the national grid.

Natural gas and natural-gas-based products, such as compressed natural gas (CNG), are used for cooking, heating, electricity generation and transport. Natural gas is the preferred fuel of choice for cooking and heating in many high-income countries. CNG is used for transport, for example in cars. Natural gas is also used for electricity generation. However, unlike coal it is often used for peak-load electricity generation. Natural gas power plants are extremely flexible, easy to start and shut down. They can therefore be adjusted to the specific electric load needs at peak time. Natural gas has the highest energy efficiency of all fossil fuels and the lowest emission factor. This means that it contributes less to global warming than coal and oil do. Gas-fired power plants usually take the form of a combined cycle gas turbine (CCGT) (sometimes also called gas turbine combined cycle plant [GTCC]) or gas turbine (GT) peak. GT peak is a very flexible gas turbine that can be switched on and off within minutes depending on peak demand. Natural gas is referred to as a transition fuel by some countries that depend heavily on it for their energy generation, such as the Netherlands and the UK. Their argument is that coal and oil should be replaced by natural gas while renewables are not yet 'ready' to take the place of fossil fuels. Yet, with the recent rapid decreases in the cost of renewables and the extensive upscaling of renewable capacity across the world, this argument is becoming increasingly less valid.

Integrated gasification combined cycle (IGCC) power plants

One state-of-the-art fossil fuel technology is the IGCC. These are also called ultra-super-critical power plants. The World Coal Association suggests that state-of-the-art coal power plants such as IGCCs have the potential to emit up to 40% less carbon dioxide than average coal power plants. They also suggest that 1% improvement in the energy efficiency of coal power plants results in 2%–3% less carbon dioxide emissions. The global average efficiency of coal-fired power plants is suggested to be around 33%, while IGCCs can have an efficiency of up to 45% (World Coal Association, n.d.). Figure 14.1 shows the average efficiency of fossil fuel power plants and their carbon dioxide emissions, in comparison to the most advanced IGCCs.

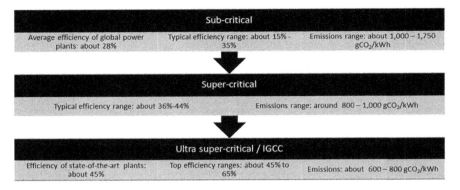

FIGURE 14.1 Efficiency and carbon dioxide emissions of fossil fuel power plants

Source: Based on IEA (2006) cited in World Coal Association (2014: 5)

Conventional steam power plants burn coal. In contrast, IGCC power plants dry the coal, pressurise it under high temperatures and thereby convert it into gas with the help of air or pure oxygen. The resulting gas mainly consists of carbon monoxide (CO) and hydrogen (H_2). The gas is cooled down and particulate matter removed. The carbon monoxide is then converted into carbon dioxide with the help of water vapour. This step is called 'shift level'. The sulfur is removed from the carbon dioxide, which is fed into the CCS system, if available. The remaining carbon dioxide is mixed with nitrogen and then used in a CCGT to generate electricity (RWE, 2014). The gas is burnt in a generator to produce electricity. For the CCS component, the remaining flue gases are used for steam generation to generate electricity. This IGCC process is also called coal gasification or coal-to-gas (RWE, 2014). Other advanced coal power plants use a process called coal-to-liquid.

Carbon capture and storage (CCS)

Carbon capture and storage (CCS) is an end-of-pipe technology for fossil fuel power plants, such as coal-fired power plants, and for industries, such as cement production and refineries. It captures the carbon dioxide emitted from the power plant or industry and then stores the carbon dioxide away from the atmosphere, often underground. This process is called carbon sequestration. The carbon can be stored in geological formations, such as saline formations, in mineral carbonates, such as depleted oil and gas fields, and in the ocean beds, or it can be used for industrial processes. CCS has been used for decades by the oil and gas industry to enhance oil and gas recovery, but it is only relatively recently that interest has developed in using this technology for carbon sequestration.

A power plant equipped with CCS technology could reduce carbon emissions to the atmosphere by approximately 80%–90% compared with a power plant without CCS technology. Nevertheless, this benefit is challenged by the high-energy use of CCS systems. The Intergovernmental Panel on Climate Change (IPCC) esti-mates that power plants equipped with CCS technology require about 10%–40% more energy than an equivalent power plant without CCS technology, while stor-age and transport of the carbon increases the energy needed to between 60% and 180% compared with conventional power plants (IPCC, 2005; Gibbins & Chalm-ers, 2008). This additional energy use produces, in turn, additional carbon emissions. There is also a range of technological and operational uncertainties related to CCS, such as carbon leakage and the long-term storage of carbon dioxide underground. Carbon leakage from the storage location into the atmosphere would compromise the opportunity of CCS to contribute to climate change mitigation (IPCC, 2005). The long-term storage of carbon dioxide poses a challenge as the time frame could be thousands, if not millions, of years. Hence, managing stored carbon would be an intergenerational operation that goes far beyond conventional governance. CCS is controversial, even in countries such as the USA and China, which are heavily reliant on coal-fired power plants. Its controversial nature stems from concerns about long-term storage and potential leakages, uncertainty related to costs and how investments in CCS influence investments in and support for other low car-bon energy options (Stephens, 2013).

There are currently 15 CCS plants operating in the world and seven under construction. This means that the number of plants has doubled in the last 10 years. According to the Global CCS Institute (2016), the total carbon dioxide capture capacity of these 22 projects is around 40 million tonnes per annum (Mtpa) (Global CCS Institute, 2016). Another concern is that CCS is an end-of-pipe technology; it therefore does not decrease the reliance on fossil fuels. It implies that humans can continue doing 'business as usual' by following a high carbon pathway and it ignores wider energy-related problems such as fossil fuel resource depletion. There are many controversies and uncertainties attached to CCS. Some countries, such as Germany, have strict CCS regulations to avoid problems of leakage and doubts about long-term storage.

Marshall (2016) argues that the rhetoric of 'clean coal' technology is a 'fantasy of technology' rather than an actual reality, as there are continuing technological challenges including regarding how to detect leaks, an absence of large-scale opera-tional prototypes and limited R&D investments in CCS by coal companies. Instead, the politics of 'clean coal' are used as a justification to extend coal deployment rather than to transition to cleaner fuels while at the same time alleviating political and social concerns about coal use (Marshall, 2016). Haszeldine (2009) suggest that various technological, economic and political hurdles still need to be overcome before CCS can play a large role in mitigating climate change.

Figure 14.2 shows a schematic overview of how CCS works.

Unconventional fossil fuels

Recently, more and more unconventional fossil fuels have been discovered such as oil shale, oil sands, tight oil, tight gas, shale gas, tar sands, coal-bed methane, etc.

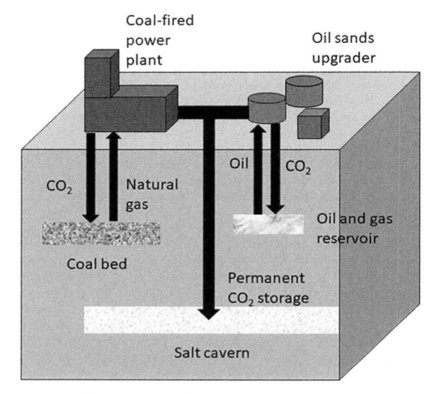

FIGURE 14.2 Schematic overview of CCS plants and carbon sequestration

Source: Author, redrawn from CeDEP, SOAS, 2015

Their exploitation is more expensive than that of conventional fossil fuels, is technically more challenging and involves a range of environmental and social issues, such as related to water pollution.

Although unconventional fossil fuel resources are more commonly exploited these days, it is estimated that even in 2050, about 85% of all the liquid fuel production will still be from conventional oil, while unconventional oil will make up about 12% and biofuels 3%. This compares to 90% conventional oil, 8% unconventional oil and 2% biofuels in 2015 (Center for Sustainable Systems, University of Michigan, 2018).

For exploiting some of these unconventional fossil fuel resources, horizontal high-volume hydraulic fracturing (HVHF) (also called fracking) is needed. Fracking extracts unconventional fossil fuels such as shale gas, tight gas, tight oil and coal-bed methane. Large amounts of water, sand and chemicals are injected at high pressure into boreholes in deep rock formations. This mining technique results in small fractures in the rock formations, which enable the gas and oil deposits to be exploited. This procedure tends to be expensive and requires a lot of energy and water. Millions of tonnes of freshwater are needed for fracking one shale gas well. There are concerns about the potential contamination of groundwater due to the use of chemicals that are needed alongside water for fracking, as well as the release of air pollutants and greenhouse gas emissions that contribute to climate change (EPA, 2012). There are also health concerns associated with fracking, particularly due to the possible contamination of groundwater from fracking. Contamination of drinking water has already happened, for example, in Pennsylvania, where a fracking well was improperly set up and maintained. There are five stages of the hydraulic fracturing water cycle that need to be critically monitored to avoid water contamination: (a) water acquisition, (b) chemical mixing, (c) well injection, (d) flowback and produced water, and (e) wastewater treatment and waste disposal (EPA, 2012). Meng (2017: 953) argues that fracking results in impacts on the 'atmosphere, hydrosphere, lithosphere and biosphere through the significant input or output of water, air, liquid or solid waste disposals, and the complex chemical components in fracking fluids.'

Some countries have banned fracking, at least temporarily, such as France and recently also Germany, others are promoting fracking, such as the USA and the UK.

A diagram of fracking is presented in Figure 14.3.

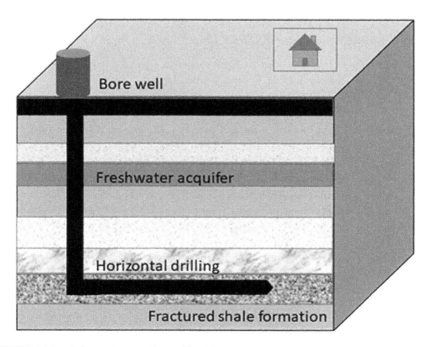

FIGURE 14.3   Schematic overview of fracking

Source: Author, redrawn from CeDEP, SOAS, 2015

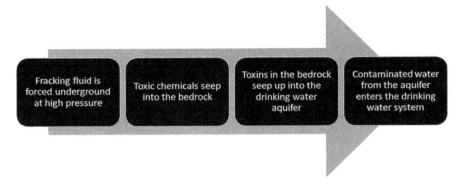

| Fracking fluid is forced underground at high pressure | Toxic chemicals seep into the bedrock | Toxins in the bedrock seep up into the drinking water aquifer | Contaminated water from the aquifer enters the drinking water system |

FIGURE 14.4   Schematic overview of the risk of groundwater contamination associated with fracking

Source: Based on information gathered from EPA (2012) and Meng (2017)

Figure 14.4 details the potential risks of contamination of groundwater through fracking.

Path dependency and carbon lock-in

Path dependency and carbon lock-in

Fossil fuel technology, including advanced fossil fuel technology, comes with two main challenges: carbon lock-in and path dependency.

Carbon lock-in can best be described as a form of inertia created by fossil fuel energy systems that results in investments, infrastructure, emission trajectories and development trajectories being locked into high carbon pathways for decades and thereby preventing a transition to climate-friendly low carbon technologies (Unruh, 2000).

Carbon lock-in plays a role as the investments and the infrastructure for energy may be tied to high carbon pathways for decades due to the long construction times and lifetimes of fossil fuel power plants, as well as the associated high investments and running costs. For example, a coal-fired power plant has a lifetime of about 20 to 40 years, or in some cases even longer. Hence, investing in the coal-fired power plant means the investments and the infrastructure will be locked into coal for the next 20 to 40 years in the absence of alternative investments and infrastructure in more sustainable energy sources. Also, fossil fuel power plants like coal-fired power stations are usually expensive and require a huge amount of investments at a specific point in time. In contrast, investing the same amount of money in renewable energy, such as wind farms, solar parks and biogas plants is usually cheaper (depending on installed capacities) and can be done at various points over time. Renewable energy also has more flexible construction times, as it only takes a few months to a couple of years to install renewable energy plants, depending on size.

A similar argument about future pathways holds for *path dependency*, in addition to a historic lock-in created by the past. Path dependency is a concept that explains how decisions about the future are determined by decisions about the past, due to past technological, economic, political, institutional, social and environmental choices.

Most of the world's systems are built to sustain a path dependency based on fossil fuels, for example urban areas that depend on roads and motorised transport, which make alternative ways of transport difficult. As most of the word's modern energy systems are based on fossil fuels, there is a high likelihood that this dependency on fossil fuels will continue into the future, if we do not intervene actively and transition towards low carbon technologies. As an example, modern societies need fossil fuels for energy generation, transport, heating, industries and so on. This is one reason why electric vehicles are still marginalised in many countries, as charging infrastructure for them is largely absent and consumers prefer cars with internal combustion engines. Another example is coal-fired power plants, which are still being built around the world, in the USA, China, Germany and elsewhere, due to path dependency, rather than transitioning to a system based on large-scale renewable energy technology such as offshore wind and concentrated solar energy technology. Changing fossil fuel path dependency would require a major shake-up of the system, mainly from a political and social perspective, as most of the renewable energy technologies are technically mature and increasingly cost effective. In this line of reasoning, Germany is planning a coal exit, albeit a very slow one until 2038. Jobs and the economy are given overriding priority in a sector that historically contributed to Germany's industrial heyday. 40 billion Euros will be made available by the German government for structural change and development projects in the former coal-producing areas, e.g. to create new employments and alternative industries (BMWI, 2019).

How to overcome carbon lock-in

Unruh (2002: 317) argues that 'industrial countries have become locked into fossil fuel-based energy systems through path-dependent processes driven by increasing returns to scale. Carbon lock-in arises through technological, organisational, social and institutional co-evolution'. He argues that overcoming carbon lock-in will not be achieved by internal forces but requires external forces. External forces can be market breakthrough or technological breakthrough, which rarely come from incumbents but rather from innovative niche players (see, for example, Tesla's innovations in electric vehicles rather than established automobile firms like BMW, Mercedes, Audi, Vauxhall, Jeep, etc.). Other options to escape carbon lock-in could be through increased public awareness of climate change and its influence on policymakers. A third option would be a climate crisis, such as the hurricane season in 2017 or the global heat wave in 2018, which raised the media attention for climate-related disasters and could in turn have an effect on policymaking. This has happened to some extent with the 2015 Paris Agreement, which was a political statement and a diplomatic success after years of political stalemate on climate policy.

The global implications of carbon lock-in

Unruh and Carrillo-Hermosilla (2006) argue that carbon lock-in and path depend-ency are not only occurring in developed countries, but also in countries that are newly industrialising, such as emerging economies that are moving away from agricultural economies to industry-based and service-based economies. Unruh and Carrillo-Hermosilla (2006) suggest that newly industrialising countries are not likely to leapfrog carbon-intensive development. Rather, they follow the develop-ment trajectories of developed countries. To some extent this has been observed throughout the recent decade as countries such as China, India and South Africa have indeed invested heavily in fossil fuel resources, mostly in coal. Many develop-ing countries still aim for the same development trajectories as the OECD coun-tries had during their industrialisation. This raises ethical issues, as poorer countries have a right to develop. Yet, over the last decade we have seen ambitious efforts by countries such as China, India, South Africa, Brazil and others to move away from carbon-intensive development and to pursue low carbon development. More recently, China, in particular, has taken the lead in investing more than any other country in the world in renewable energy, installing more renewable energy capac-ity than any other country in the world (IEA, 2019b), exporting its low carbon technology overseas and investing in low carbon infrastructure in many other countries around the world and playing a leadership role in global climate policy.

ENERGY PLANNING SCENARIOS

We now look at two hypothetical scenarios to apply the theories of path dependency and carbon lock-in.

Country A decides to invest US$1.5 billion in a new 500 MW coal-fired power plant in 2020, which takes 4 years to build and has a lifetime of 35 years or more.

Country B decides to invest a similar amount of money, US$1.35 billion, in a 200 MW wind farm in 2020 (US$450 million in total), which takes about a year to build, four solar parks at 25 MW each to be built in 2020 (US$100 mil-lion in total), which take a few months to build, four biogas power plants pow-ered by agricultural/livestock residues from nearby farms at 25 MW, each to be built in 2030, which each take a year to build (for US$800 million in total). The lifetimes of these power plants are also 35 years or more.

In 2022, after two construction years, country A has 0 MW new installed capacity as the coal-fired power plant is still being built but it cannot invest in other power plants as the investments have already been committed. Country B has 300 MW new installed capacity from wind and solar in 2022 and plans to invest the remaining money into the biogas plants in 2035.

Country A is dependent on coal imports from another country, has high fuel costs in addition to operation and maintenance costs and has a low energy security due to relying on one fuel only. It will be bound to these investments for the next 35 years as all energy expenses have been made until the end of this planning period.

Country B uses indigenous, renewable energy resources, has some operation and maintenance costs but no fuel costs, and has a higher energy security due to its diversified energy portfolio. It is not dependent on imports from other countries. In the event of technical failures or other problems with one of the power plants, there are a variety of other plants to rely on.

In 2030, country A reduces the output of the coal-fired power plant from 500 MW to 400 MW due to declines in industrial productivity in the region and hence lower energy demand. Yet, the investment for the full 500 MW has been made. Hence, there is a carbon lock-in concerning both investments and infrastructure.

Also, by 2030, country B decides to build only two of the biogas power plants, adding a further 50 MW to the grid by 2031, but de-commissioning the contracts for the other two plants (letting the contractors know well in advance to avoid contract issues). The additional 50 MW are not needed, as a similar capacity is being generated by distributed energy from the many rooftop solar photovoltaic (PV) systems that are feeding electricity into the grid across the country. The additional investment for the 50 MW is therefore saved for the next planning period, hence freeing some cash for future investments in the energy sector. Also, since the prices of solar and wind energy became more competitive around 2015, the investments in renewable energy turn out to be more cost effective than investing in coal- or gas-fired power plants.

Country A and country B are signatories to the United Nations Framework Convention on Climate Change (UNFCCC) and the Paris Agreement. They have Nationally Determined Contributions (NDCs) in place which outline how they intend to reduce emissions between 2020 and 2030, as well as on a long-term basis until 2050. Country B has achieved their emission reduction targets by 2030 due to diversifying their energy portfolio and achieving a partial transition from fossil fuels to wind, solar and biogas (some fossil fuel plants are still operating in the country though). Country A has not managed to reduce its emissions, due to its high dependency on coal. It now needs to develop plans to reduce emissions until 2050 to avoid penalties such as needing to buy emission reduction credits from other countries.

Exercises

1   Make a list of the latest state-of-the-art fossil fuel technology. Briefly explain how each of these technologies work, as well as their advantages and disadvantages.

2   Think of a case of carbon lock-in and path dependency for a country of your choice.

References

BMWI German Federal Ministry for Economic Affairs and Energy. (2019) *Kohleausstieg und Strukturwandel (Coal Exit and Structural Change)*. Available from: https://www.bmwi.de/Redaktion/DE/Artikel/Wirtschaft/kohleausstieg-und-strukturwandel.html

Center for Sustainable Systems, University of Michigan. 2018. "Unconventional Fossil Fuels Factsheet." Pub. No. CSS13–19.

EC. (2018) *Energy and Climate Targets*. Available from: https://ec.europa.eu/clima/policies/strategies/2050_en

EPA. (2012) *Study of the Potential Impacts of Hydraulic Fracturing on Drinking Water Resources*. US Environmental Protection Agency (EPA). Progress Report. Available from: http://www2.epa.gov/sites/production/files/documents/hf-report20121214.pdf

Gibbins, J. & Chalmers, H. (2008) Carbon capture and storage. *Energy Policy*, 36 (12), 4317–4322.

Global CCS Institute (2016) *Large Scale CCS Projects*. Global CCS Institute. Available from: www.globalccsinstitute.com/projects/large-scale-ccs-projects

Haszeldine, R.S. (2009) Carbon capture and storage: How green can black be? *Science*, 325 (5948), 1647–1652.

IEA (2006) *Focus on Clean Coal*. OECD and IEA Working Paper. November 2006. Available from: www.iea.org/publications/freepublications/publication/focus_on_coal.pdf

IEA. (2019a) *Sweden Is a Leader in the Energy Transition, According to Latest IEA Country Review*. Available from: https://www.iea.org/newsroom/news/2019/april/sweden-is-a-leader-in-the-energy-transition-according-to-latest-iea-country-revi.html

IEA. (2019b) *Statistics*. Paris, International Energy Agency (IEA), OECD/IEA. Available from: www.iea.org/statistics/

IPCC. (2005) Summary for policymakers. In: *IPCC Special Report on Carbon Dioxide Capture and Storage*. Prepared by Working Group III of the Intergovernmental Panel on Climate Change. [Metz, B., O. Davidson, H.C. de Coninck, M. Loos, and L.A. Meyer (Eds.)]. Cambridge University Press, Cambridge, United Kingdom and New York, NY, USA, 442 pp. Available from: www.ipcc.ch/pdf/special-reports/srccs/srccs_summaryforpolicymakers.pdf

Marshall, J.P. (2016) Disordering fantasies of coal and technology: Carbon capture and storage in Australia. *Energy Policy*, 99 (12), 288–298.

Meng, Q. (2017) The impacts of fracking on the environment: A total environmental study paradigm. *Science of The Total Environment*, 580 (2), 953–957.

RWE. (2014) *IGCC/CCS Power Plant*. RWE. Available from: www.rwe.com/web/cms/en/2688/rwe/innovation/projects-technologies/power-generation/fossil-fired-power-plants/igcc-ccs-power-plant/

SOAS. (2015) *Distance-Learning Module Energy & Development*. London, SOAS.

Stephens, J. (2013) Carbon capture and storage (CCS) in the USA. In: Urban, F. & Nordensvärd, J. (Eds.) *Low Carbon Development: Key Issues*. Oxon, Earthscan, Routledge. pp. 297–307.

Unruh, G. (2000) Understanding carbon lock-in. *Energy Policy*, 28 (12), 817–830.

Unruh, G. (2002) Escaping carbon lock-in. *Energy Policy*, 30 (4), 317–325.

Unruh, G. & Carrillo-Hermosilla, J. (2006) Globalizing carbon lock-in. *Energy Policy*, 34 (10), 1185–1197.

World Coal Association (n.d.) *High Efficiency Low Emission Coal*. Available from: www.worldcoal.org/reducing-co2-emissions/high-efficiency-low-emission-coal

World Coal Association (2014) *A Global Platform for Accelerating Coal Efficiency*. Platform for Accelerating Coal Efficiency (PACE) Concept Paper. London, World Coal Association. Available from: www.worldcoal.org/sites/default/files/resources_files/pace_concept_paper%2809_01_2015%29.pdf

15

TECHNOLOGY FOR ENERGY AND DEVELOPMENT

Low carbon energy and energy efficiency

Technological advances: energy efficiency and low carbon energy

Energy efficiency, smart grids and combined heat and power

Energy efficiency

Energy efficiency is an important, though often overlooked, way of reducing the costs and the environmental impacts of energy. Goldemberg and Lucon (2009) describe how energy efficiency can be advantageous:

- Energy efficiency is the most cost-effective way of mitigating climate change.
- It leads to a saving of energy resources, usually referred to as energy conservation.
- This means that energy resources are saved and energy security (expressed as security of supply) increases.
- This can lead to an increase in productivity and industrial competitiveness, which has economic benefits.
- Energy efficiency contributes to reducing the emissions of greenhouse gases and other pollutants, thereby mitigating climate change and reducing air pollution.

Goldemberg and Lucon (2009) indicate that without energy efficiency measures applied in the OECD between 1973 and 1998, energy consumption would have been 50% higher. They also suggest that developing countries could save up to 65% of their energy consumption between 2006 and 2026 by adopting energy efficiency measures.

As discussed in Chapter 12, marginal abatement cost (MAC) curves are greenhouse gas (GHG) abatement cost curves that are used to calculate the annual carbon abatement potential for technological interventions and the costs of mitigation. These cost curves indicate that energy efficiency measures, such as installing smart meters and energy-efficient lighting, are the cheapest option for reducing carbon dioxide emissions and thereby reducing the GHG emissions leading to climate change. Yet, smart meters are only monitoring tools and therefore they are only useful if the users know how to reduce their energy use, e.g. by turning off devices they do not need such as lighting or TV sets that are on stand-by, switching from standard light bulbs to energy-efficient light bulbs etc. The effective use of smart meters therefore involves an element of learning and behavioural change. Casillas & Kammen (2010) discuss how energy efficient measures can be applied to rural energy systems in Latin America. This is, however, not a trend limited to rural areas or to the developing world; instead energy efficiency measures are the most cost-effective way of reducing GHG emissions around the world, in urban and in rural settings, in high-, middle- and low-income countries. This is the conclusion that leading scholars, consultancies and international institutions such as the International Energy Agency (IEA) have come to with regard to the most cost-effective way of mitigating climate change (Diesendorf, 2014). MACs (such as presented in Chapter 12) show that energy-efficient technology and processes are the most cost-effective way of reducing GHG emissions, whereas carbon capture and storage (CCS) is the most expensive option for reducing GHG emissions.

The main types of energy-efficient technology and processes can be found in buildings and industry:

- energy-efficient building design: passive houses, houses with solar photovoltaic (PV) and/or solar hot water heaters, geothermal heat pump, energy-efficient cavity insulation, loft insulation, floor insulation, heat recovery ventilator (HRV)
- energy-efficient appliances for buildings: energy-efficient light bulbs, light emitting diodes (LEDs), electronic equipment, refrigeration, home appliances, etc.
- energy-efficient processes and appliances for industry: a wide range of processes and appliances including steel manufacturing, concrete manufacturing, aluminium smelting, etc.

Table 15.1 shows eight kinds of energy 'wastes' in industrial processes and how they can be reduced through energy-efficient processes. This is not only beneficial for the environment, but also saves firms unnecessary costs.

Energy-efficient buildings are becoming more popular. The EU's Energy Performance of Buildings Directive requires that all buildings that are being sold, built or rented must have an Energy Performance Certificate (EPC). In addition, larger public buildings over 500 m² (250 m² from July 2015 onwards) must display a

TABLE 15.1 Energy efficiency measures that involve eight ways of reducing waste

Kind of waste or unused potential	Definition	Example
Over-production	Producing excess energy – input energy that is unused	Venting excess steam
Waiting	Consuming energy while production is stopped	Laser welding line on standby still consumes 40% of maximum energy
Transportation	Inefficient transport of energy	Leaks and heat radiation in steam networks
Over-specification	Process energy consumption higher than necessary	Blast furnace is operating at 1100 degrees C rather than the required 1000 degrees C
Inventory	Stored goods use/lost energy	Crude steel cools in storage, is then reheated for rolling
Re-work/scrap	Insufficient re-integration in upstream process when quality is inadequate	Re-drying polymer lines that did not get coagulated in drying process
Motion	Energy-inefficient processes	Excess oxygen in steam boiler
Employee potential	Failure to use people's potential to identify and prevent energy waste	Employees not involved in developing energy-saving initiatives

Source: Amended from McKinsey and Company (2010a: 32)

Display Energy Certificate (DEC), and all air-conditioning systems over 12 kW and heating systems over 20 kW must be regularly inspected by an energy assessor. Germany has particularly high energy-efficiency standards for buildings. It has many new-build properties of a very high standard. A type of house called the passive house is particularly energy efficient. A passive house produces more energy than it requires, for example, due to its energy-efficient design and its rooftop solar PV systems (that enable its owners to feed electricity back into the grid and thereby earn money through the feed-in tariff) or its solar hot water heaters. The energy consumption and carbon dioxide emissions from various types of houses is shown in Figure 15.1.

Energy efficiency has several potentials: the theoretical potential (that could theoretically be achieved), the technical potential (that is technically possible), the economic potential (that is technically possible and economically feasible) and the market potential (that is technically possible, economically feasible and commercially available). Barriers to energy efficiency include a lack of an international mandatory standard for energy efficiency, weak legislation, financial issues and issues related to accountability (e.g. tenants living in a house may wish to have a more energy-efficient house to reduce energy bills, but as they do not own the house they are unlikely to invest in energy efficiency, whereas the house owner may not care as (s)he does not pay the bills).

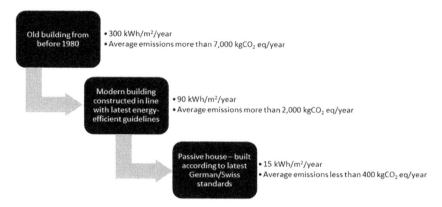

FIGURE 15.1 Comparison of different types of buildings and their energy consumption and CO_2 emissions

Source: Based on information from UNEP/Grid-Arendal (2009)

Another concept related to this debate is the energy conservation, which means a reduction of energy demand. For example, not all lighting in an office may be needed, hence turning off some lights would save energy without an improvement in energy efficiency (Diesendorf, 2014). Energy conservation is a key strategy for the Chinese government, as China's energy demand puts a strain on domestic natural resources. It has received less attention in OECD countries where the focus is often on economic growth and improving lifestyles, rather than achieving this while reducing energy demand and encouraging energy saving.

Both energy efficiency and energy conservation are challenged by the rebound effect. The rebound effect is a phenomenon whereby people and/or institutions save money by saving energy and then spend the money on other energy-consuming activities. For example, someone who saved money by driving a more fuel-efficient car may use the money saved to fly on a family holiday, or an office that saves money by installing energy-efficient light bulbs may use the money saved to buy a new printer. The energy that is saved by the first energy-consuming activity may be spent again by the second energy-consuming activity. However, the rebound effect is estimated at only around 8%–10% of gross domestic product (GDP) (Diesendorf, 2014). In addition, not everyone will spend the saved money on energy-consuming activities. Energy efficiency technology and energy conservation enables people, households and firms to make real money savings, which is particularly important for people and households that have low incomes or those who are struggling paying their bills.

Smart grids

Smart grids are modern grids that use information and communication technology to improve the efficiency, cost effectiveness, sustainability, flexibility and responsiveness of the grid. Smart meters are one part of smart grids, which provide detailed information about energy use and behaviour. They can provide control over the grid to both the grid operator and the consumer (Tawney, 2013).

Smart grids have the potential to deal better with variable or intermittent renewable energy, as well as decentralised electricity. This allows for the large-scale introduction of renewable energy. Smart grids enable bidirectional energy flow, meaning that the electricity can both flow from the grid to the user (e.g. the grid providing electricity to a household) and from the user to the grid (e.g. a household providing electricity to the grid from its rooftop PV system). Electricity use is not stable during the day or the year, but varies according to customer demand, day and night, winter and summer. Smart grids can respond to peak loads and increase pricing when there is excessive electricity use to reduce the pressure on the electricity system, whereas a higher electricity demand during times of lower electricity use can be encouraged at cheaper costs. Smart grids can also warn electricity users to lower their demand during peak times (while additional power plants are starting up), for example, by sending signals to TV sets to turn to standby. This can help to reduce peaks in energy use and avoid blackouts and load shedding. This is called peak curtailment.

Combined heat and power

Combined heat and power (CHP), also known as co-generation, is a process by which a power plant generates electricity and useful heat simultaneously. Combined cooling, heat and power (CCHP) (also known as tri-generation or poly-generation) is a process by which a power plant generates cooling, electricity and useful heat simultaneously. CHP and CCHP is an energy-efficient form of energy generation. It is often used for district heating, for example, in many European countries (including in major cities such as Stockholm, Berlin and London), the USA and Canada (including in New York and Montreal), China and Japan. Iceland is famous for its large-scale district heating CHP from geothermal energy. CHP, and particularly district heating CHP, is one of the cheapest and most energy-efficient ways of cutting carbon emissions.

Renewable energy

This section briefly mentions the key renewable energy sources and their technologies. It then goes on to discuss current advances in these technologies.

Renewable energy technology

Renewable energy comes from renewable natural resources, such as sunlight, wind, water, tides, geothermal heat and biomass. Unlike fossil fuels and nuclear energy, which are finite and depletable, these energy resources are renewable and non-depletable. Renewable energy has a large global potential. The World Energy

Council estimates that the theoretical potential is 370 PWh/year for solar energy, 315 PWh/year for primary biomass, 96 PWh/year for wind energy and 41 PWh/year for hydropower. Nevertheless, the technical and economic potential is lower due to variations in land availability and financial competition with fossil fuels (WEC, 2007). About 24% of the global electricity consumption came from renewable energy in 2018, mainly from hydropower, but also from wind, solar and biomass (IEA, 2019). The most widely used and commercialised renewable energy technologies are wind turbines, PV panels and hydropower technology.

Wind energy technology converts wind into electric power. It thereby converts kinetic energy into mechanical energy. It has been used for several centuries in the form of windmills. Modern wind energy technology such as the wind turbine, has seen a rapid growth within the last few decades. After hydropower, wind energy is the most widely used and commercialised renewable energy technology. Wind energy has seen a rapid global increase between 1996, when only 6.1 gigawatts (GW) were installed, and late 2018, when about 597 GW were installed, of which 96% onshore and 4% offshore (WWEA, 2019; GWEC, 2019). The countries with the largest installed wind energy capacity are the USA and China, followed distantly by Germany, Spain, India and the UK (WEC, 2017). Figure 15.2 shows the global installed wind capacity between 2001 and 2018, which has been steadily increasing.

Wind turbines can range from small turbines, which are usually in the range of kilowatts (kW) and located onshore – on land – to large offshore turbines, which are in the range of megawatts (MW) and located offshore – in the sea. One modern large wind turbine, such as Vestas' V164 9.5 MW wind turbine, Siemens SWT-8.0–167-DD wind turbine with a generation capacity of 8 MW and Enercon's E-126 turbine, which has a generation capacity of 7.58 MW can power up to 5,000 households that have an average European energy demand or even more. Wind turbines can be connected to the central grid or used in mini-grids or as decentralised

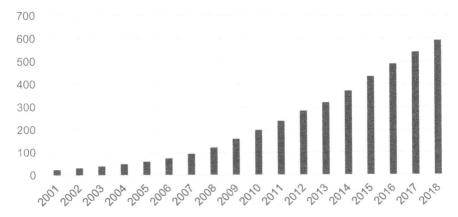

FIGURE 15.2 Global cumulative capacity of wind energy, measured in GW

Source: Data from GWEC, 2019

stand-alone systems. Lifecycle analysis shows that the GHG emissions from wind energy technology are significantly below those of fossil fuel plants (Gagnon et al., 2002; Evans et al., 2008).

Solar energy technology converts solar radiation into electric power or heat. There are various forms of solar energy technology, including PV, solar thermal technology (like solar water heaters), concentrated solar power (CSP), solar lamps and solar cookers (used in some developing countries for cooking). Solar water heaters can often be found on rooftops of individual houses or buildings, whereas CSP is often used on a large scale to replace fossil fuel power plants. Solar PVs can be found both on individual buildings and on a large scale to replace power plants. The most widespread solar energy technology is solar PV, which converts solar radiation into electricity by using semi-conductor materials and their photovoltaic effect. These materials include various forms of silicon (mono-crystalline, policy-crystalline and amorphous) as well as gallium sulphides. Solar PVs can be connected to the central grid or used in mini-grids or as decentralised stand-alone systems. Lifecycle analysis shows that the GHG emissions from solar energy technologies are significantly below those of fossil fuel plants (Gagnon et al., 2002; Evans et al., 2008).

Figure 15.3 shows the global installed solar PV capacity between 2007 and late 2017, which has been steadily increasing. In late 2017, the global installed capacity for solar PV was nearly 405 GW compared with 10 GW in 2007 (Global Solar Council, 2018). The regions with the largest installed solar PV capacity are Asia-Pacific, particularly China and Japan, followed by Europe, particularly Germany but also Italy, the UK, France and Spain, and North America, especially the USA. China today operates about one third of global solar installed capacity (Global Solar Council, 2018).

Hydropower technology converts the energy of falling water into electricity or mechanical energy. It has been used for several centuries for powering mills and

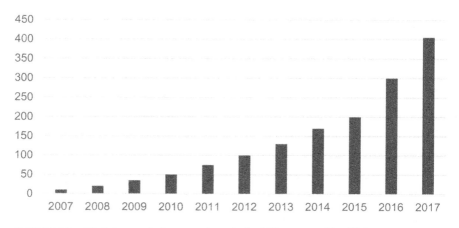

FIGURE 15.3 Global cumulative capacity of solar PV, measured in GW

Source: Global Solar Council, 2018

industries. Nevertheless, since the 20th century, it has mainly been used to generate electricity. Hydropower is the most widely used and commercialised renewable energy technology by far. Some countries, such as Norway, Nepal and Mozambique, depend almost entirely on hydropower for electricity generation. In 2017 it supplied close to 17% of global electricity with an installed capacity of about 1.27 terawatts (TW or about 1270 GW), far more than any other renewable energy source. This hydro capacity generated about 4,200 terawatt hours (TWh) of electricity in 2017, and estimates suggest it mitigated about 4 billion tonnes of greenhouse gases, such as carbon dioxide (IHA, 2019). The highest share of hydropower capacity is installed in China (nearly a third of global installed capacity), followed at some distance by the USA, Brazil, Canada, Japan, India and Russia (IHA, 2019).

There is a distinction between large hydropower and small hydropower. Large hydropower refers to hydropower technology that has a generation capacity of more than 10 MW and often involves the construction of large dams and reservoirs. Dams have been the focus of much controversy in the 1990s and early 2000s due to their potential for severe environmental and social destruction. With the introduction of stricter policies for environmental impact assessments (EIA), resettlements and compensation of resettled people, some of the controversy has subsided, although major shortcomings are still being reported, particularly in Southeast Asia and Africa. Small hydropower refers to hydropower technology that has a generation capacity of below 10 MW and is predominantly based on river run-off (ESHA, 2009). It is recognised as having low environmental and social impact compared with large hydropower schemes. Pumped storage is a type of energy storage using hydropower that had a capacity of over 150 GW in 2018 (IHA, 2019). Lifecycle analysis shows that the greenhouse gas emissions from both small and large hydropower technologies are significantly below those of fossil fuel plants (Gagnon et al., 2002; Evans et al., 2008).

Modern biomass such as biofuels and biogas is derived from organic material and is used for transport purposes (bioethanol and biodiesel), cooking (biogas) or electricity generation (biogas, waste-to-energy, wood chips, etc.). Biogas and waste-to-energy are usually derived from agricultural or livestock residues such as poultry litter, which is then gasified and incinerated. This can take the form of large biomass power plants that feed electricity into the grid, for example, in the UK and in Germany, or small-scale decentralised biogas production that is mainly used locally, for example, in rural India. Biogas is used in several South Asian countries as a cleaner, more efficient and healthier cooking option than traditional biomass.

Biofuels are much more controversial. These include first-generation, second-generation and third-generation biofuels. First-generation or conventional biofuels are usually based on edible biomass-based starch, sugar or vegetable oil. This means these fuels are usually based on food products such as corn, wheat or other cereals, cassava and sugar beets that are used to make bioethanol. Brazil, for example, has had very successful bioethanol programmes in place for many decades, making Brazil the global forerunner in bioethanol production with the world's largest vehicle fleet operating on bioethanol. As Brazil's bioethanol comes from sugar cane, it does not threaten food security (Dequech-Neto & Heiss, 2013). Soy, jatropha and palm oil are used for making biodiesel. Second-generation biofuels are biofuels that are not edible, but they are based on feedstock. This can include, for example, municipal waste. Third-generation biofuels or advanced/unconventional biofuels usually do not depend on food or feedstock products but can be derived from algae, cellulose and other forms of plant biomass, which solves problems related to food security but makes it harder to extract fuel (Goldemberg & Lucon, 2009). Many countries are increasingly producing biofuels for transport. Unfortunately, it has been reported that this may negatively affect food security. The food security issue arises due to the fact that land used for biofuel production cannot be used for food production; hence, there is a conflict between land for biofuels and land for food production (Rathmann et al., 2010).

The following section discusses current advances in renewable energy innovation. We focus particularly on large-scale wind turbines, onshore and offshore, and CSP.

Large wind turbines: onshore and offshore

A wind turbine is a piece of engineering excellence as it is composed of about 10,000 parts and requires integration with the grid. Cutting-edge research in wind energy technology means that today more large wind turbines in the multi-megawatt range are developed and commercialised (or diffused, as innovation studies phrase it). Research and development (R&D) into wind turbines of 10 MW or larger is also increasing. For several years, the E-126 wind turbine, developed by the German wind energy firm Enercon, was the world's largest wind turbine with a generation capacity of 7.58 MW. It has the capacity to power 5,000 households, or an entire village, with an average German electricity demand (Enercon, 2007). In areas where the electricity demand is lower, it could power even more households, for example in rural India, where household electricity demand can be up to 10 times lower than in the EU. Enercon's powerful E-126 is situated onshore (on land). Siemens later built the largest offshore wind turbine, the SWT-8.0–167-DD, which is powered by a direct drive. The Danish wind energy firm Vestas, which is leading the global wind energy market, developed an even larger wind turbine for offshore use at sea. The V164 has a generation capacity of 8 MW and was later upgraded to 9.5 MW (Vestas, 2019). Offshore wind energy is currently experiencing a boom. Leading offshore wind energy nations such as Denmark and the UK have been operating in this sector for more than two decades and have gained considerable knowledge in how to perfect offshore logistical operations as well as the offshore technologies. Offshore business models require innovative approaches as the scale, size and weight of offshore turbines and installing them at sea is challenging, particularly in deep waters and rough weather. However, it is not only the Northern European offshore sector that is rapidly growing. Other countries in Europe, North America and Asia are increasingly investing in R&D for offshore wind turbines and operations. This is partly driven by the need to install turbines out of sight where people object less to them, due to the so-called NIMBY effect (Not in My Backyard) and to reduce the need for land for wind farms.

In Germany, as well as in India and China, wind energy experts talk about the 'second spring' for wind energy, thanks to the development and diffusion of wind turbines that are made for low wind speed areas. This means that even areas that do not have a high wind energy potential due to their low wind speeds can exploit wind energy, at a much higher energy efficiency and cost effectiveness than before. The costs for wind energy have reduced dramatically in the last few decades, as the technology is mature and utilised on a large scale in many countries.

There are currently two different technological trajectories for wind turbines: turbines with gears, such as those produced by Vestas and a wide range of other producers, and gearless turbines that operate by using a direct drive. The gearbox design is the classic design used for many decades by most wind turbine producers. The direct drive technology was pioneered by Enercon. However, today it is also used by other manufacturers such as General Electric (GE), Goldwind, Siemens, Vensys, etc. Different types of gearless turbines use different types of direct drives, for example the electromagnetic direct drive used by Enercon and the permanent magnetic direct drive (PMDD) used by Goldwind and Vensys. Turbines with direct drives are reported to have lower operation and maintenance (O&M) requirements and are therefore more reliable; however, they are more expensive than gear-driven turbines and some of them, such as the PMDD, require access to rare earths (Lema et al., 2014).

Concentrated Solar Power (CSP)

Recent advances in solar energy include utility-scale solar energy such as solar farms/parks. New technological advances also include improved design and more advanced materials for PV panels and systems, such as thin-film PV. A relatively new solar technology that is becoming increasingly popular due to its utility-scale size and its alternative to fossil fuel power plants is CSP, also known as concentrating solar power or concentrated solar thermal. CSP is a technology that converts concentrated sunlight directly to electricity due to the photovoltaic effect. CSP uses mirrors or lenses to concentrate a large area of sunlight onto a small area. This concentrated light is converted to heat, which drives a steam engine that is connected to an electric generator. There are various types of CSP plants, although parabolic plants using parabolic reflectors are by far the leading technology. Waste heat from the generation of CSP electricity can also be used for the desalination of seawater and to help to reduce water shortages. New innovation in CSP has enabled energy to be stored for up to 15 hours, which means that it can be stored overnight and when the sky is cloudy, to enable a stable and reliable electricity supply that can be used as a back-up capacity to cover for fluctuating or intermittent wind or solar PV generation when needed (Desertec, 2014).

CSP is a relatively young technology that is not as mature as wind or hydropower technology, but as Figure 15.4 shows, global installed capacity is rapidly increasing and reached 5.5 GW in late 2018 (REN21, 2019). Large-scale diffusion makes the technology increasingly cost effective. The CSP sector is being led by Spain, which has the world's largest capacity of CSP, followed by the USA. The world's largest

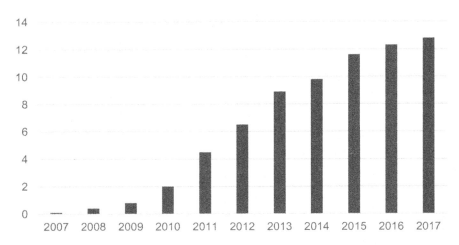

FIGURE 15.4 Global cumulative capacity of concentrated solar power (CSP) in GWh

Source: REN21, 2019

CSP plant is currently the Ouarzazate Solar Power Station in Noor, Morocco, with a generation capacity of 510 MW. Other large CSP stations can be found in the USA (particularly in California), Spain, India and South Africa. The Desertec Foundation aims to build a network of renewable energy, particularly CSP plants, and grid infrastructure in the arid areas of Northern Africa to improve energy access in these regions and to potentially enable electricity trade from Northern Africa to Europe. A large problem here is geopolitics as Northern Africa has seen increasing instability in recent years. Figure 15.4 shows growing cumulative capacity in CSP worldwide. While the sector is rapidly developing, the overall contribution to global energy supply is still rather small, e.g. comparing the 4,200 TWh of hydroelectricity generated in 2017 versus the 12.8 GWh electricity generated by CSP in the same year. Yet, increasing investments and increasing capacity additions show that CSP's importance is growing. It also has the opportunity to contribute to regions that have abundant sunshine but are troubled by energy poverty, such as across Africa, developing Asia and Latin America. Also, its large-scale character poses realistic opportunities for replacing coal-fired power plants of similar size.

Bioenergy with carbon capture and storage BECCS

Bioenergy with carbon capture and storage (BECCS) is a relatively new and yet under-developed carbon-negative technology that can be used for generating both electricity and heat. It received a lot of attention recently as it became clear that the below 2 degree goal of the Paris Agreement can only realistically be achieved using carbon-negative technologies such as BECCS.

Raftery et al. (2017) argue that it is unlikely to limit global warming to less than 2 °C. Raftery et al. (2017) suggest that the aspirational goal of limiting global warming to 1.5 °C compared to pre-industrial levels may only have a 1% chance of succeeding, while keeping global temperature rise at or below 2 °C by 2100 only has a 5% chance of succeeding. In contrary, Millar et al. (2017) argue that it will be difficult, but not impossible, to limit global warming to 1.5 °C. This will require the implementation of the pledges made for 2030 under the Paris Agreement on climate change and it will need deep and rapid mitigation actions, which are partly relying on new technology that is currently still in R&D. It is an underlying assumption in many of the scenarios developed through integrated assessment models (IAMs) that low carbon energy technology alone will unlikely be sufficient in avoiding dangerous climate change and that there is a need for technologies such as BECCS (Kato et al., 2017, Mander et al., 2017). Mander et al. (2017: 6036) argue, 'BECCS is a central, but often hidden element of many of the modelling work that underpins climate policy from the global to the national scale'.

Fridahl and Lehtveer (2018) argue that while globally BECCS is considered as important, at the national level there are various obstacles, such as limited investment, lack of policy incentives and issues around social acceptance. Most investments in BECCS have so far been in Europe and North America, for example in Sweden, Norway, the Netherlands and the USA.

How BECCS works: First, the biomass is grown in forests. Second, the biomass such as from forest products is delivered to the power plant or to the combined heat and power plant by rail, road or boat. Third, the biomass is converted into electricity and/or heat. Fourth, CO_2 is separated from the flue gases post-combustion and compressed into liquid form. Fifth, CO_2 is transported to a permanent storage location. This is often done by boat, as the storage locations are often located off-shore in depleted oil and gas fields in the seabed. Other options for storage are deep saline formations. Finally, the CO_2 is being stored underground. The cycle is then repeated by the repeated growing of biomass, such as in forests, which capture carbon as part of their natural photosynthesis process, which is then later captured and sequestrated. See Figure 15.5 for a schematic view of BECCS. Other approaches include BECCS at bioethanol refinery plants, landfill gas combustion and municipal solid waste incineration (Pour et al., 2017).

Most BECCS projects are currently still in the R&D phase, with pilot projects and demonstration projects as a next step and commercialisation planned within the next 20 years. Major projects are, for example, the Illinois Carbon Capture and Storage (IL-CCS) in the USA, which aims at storing 1 million tonnes CO_2 per year in the Mount Simon Sandstone, one of the largest saline aquifers worldwide (US Department of Energy, 2017). Stockholm Exergi's BECCS project in Sweden even aims at mitigating more than 2 million tonnes CO_2 per year that would be deposited in depleted oil and gas fields offshore (Stockholm Exergi, 2019). Levihn (2017) suggests that in Stockholm, Sweden, the costs for BECCS would be about 700–1,000 SEK per ton, hence about 70–110 USD or even cheaper if CO_2 storage from BECCS would be combined with CO_2 storage from fossil-based CCS.

There are several drawbacks and limitations to BECCS: This technology is limited to countries and regions that are rich in biomass, such as those having abundant

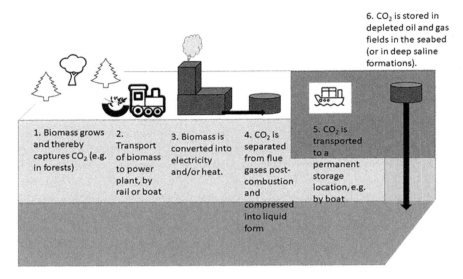

6. CO$_2$ is stored in depleted oil and gas fields in the seabed (or in deep saline formations).

1. Biomass grows and thereby captures CO$_2$ (e.g. in forests)

2. Transport of biomass to power plant, by rail or boat

3. Biomass is converted into electricity and/or heat.

4. CO$_2$ is separated from flue gases post-combustion and compressed into liquid form

5. CO$_2$ is transported to a permanent storage location, e.g. by boat

FIGURE 15.5   Schematic overview of BECCS

Source: Based on information from Stockholm Exergi, 2019

forests. Yet, to avoid BECCS contributing to deforestation, it needs to be ensured that no primary or virgin forest is deforested and that the biomass comes from rest products from forest-related industries, such as the residues from sawmills, pulp mills and logging. Using BECCS at bioethanol refinery plants, landfill gas combustion and municipal solid waste incineration can reduce the pressure on biomass. Similar to CCS, considerations need to be made regarding potential leakage and long-term governance of CO$_2$ underground. In addition, Mintenig et al. (2017) critically discuss that there is a possibility that carbon-negative technologies such as BECCS postpone climate mitigation efforts into the future and that it reduces the immediate pressure to close down fossil fuel plants.

Nuclear energy

Nuclear energy technology refers to power plants that use nuclear fission to generate electricity and heat. At a global level, there were 390 GW of installed nuclear capacity in 2017 (IEA, 2019). The countries with the largest installed capacity of nuclear power are the USA, France, China, Japan (with very reduced capacity due to the Fukushima nuclear accident in 2011), Russia, South Korea, Canada and Ukraine. Other countries with considerable nuclear operating capacity are Germany, the UK, Sweden, Spain, Belgium and Taiwan (WNA, 2019). China plans to more than double its nuclear generation capacity within the next decade, while Germany intends to completely phase out its nuclear power by 2022 and replace it with renewable energy and coal (although a coal exit is also planned until 2038).

While nuclear energy is a low carbon energy source, which produces less carbon and other GHG emissions than fossil fuels, it is fiercely debated due to severe health, safety and environmental implications. Two key controversies were exemplified in the nuclear disasters in Chernobyl, Ukraine (part of the former USSR), in 1986, and in Fukushima, Japan, in 2011. Even today, more than 25 years after the nuclear accident in Chernobyl, the affected area – including its water, soil, flora and fauna – is heavily contaminated. The WHO argues that about 50 people have died and another 4,000 people could eventually die from radiation exposure following the Chernobyl accident (WHO, 2005). In 2011, the nuclear power plant Fukushima was hit by an earthquake scale 9, which was followed by floods caused by a tsunami. This resulted in a nuclear meltdown in several of the reactors. While a few workers died after the accident, the extent of deaths and illnesses caused by the accident is not fully established, although estimates range between 100 and 1,000 expected cancer deaths alone (Caracappa, 2011; Von Hippel, 2011). Following Fukushima, recommendations were made for the nuclear energy sector to better take into account disaster risks, for example by improving siting procedures and increasing the robustness and design of nuclear reactors (Urban & Mitchell, 2011). Another problem with nuclear energy is that it depends on uranium resources. There is a heated debate whether uranium will become a rare and near-depleted resource any time soon, similar to oil, or whether uranium resources will remain abundant for hundreds or even thousands of years to come.

An innovation in nuclear energy technology is nuclear fusion, which is a nuclear reaction involving a collision of at least two atomic nuclei at high speed, which results in the formation of a new type of atomic nucleus. Nuclear fusion has advantages over nuclear fission, because it has greater energy efficiency, is easier to contain and shut off and has less radioactive waste; it is also more difficult to produce nuclear weapons from nuclear fusion material compared with nuclear fission. In short, it is a nuclear technology that is claimed to be less polluting and less dangerous. Nevertheless, there are still high environmental, health and safety risks associated with nuclear power plants, uranium is a finite resource and nuclear power plants as well as their R&D are very expensive. Nuclear fusion technology has been pursued since the 1950s and large sums of money have been invested in this technology; however, the technology is still in the R&D phase and no commercial nuclear fusion plants exist yet. Some argue that it may take only one more decade to build commercially viable nuclear fusion power plants; however, this has been claimed for several decades (Atherton, 2013).

Innovation, technology transfer and technology cooperation

Innovation and low carbon energy technology

Research shows that renewable and low carbon energy transitions are highly important climate change mitigation options (e.g. Urban, 2014, 2018). Access to low carbon energy technology innovation is crucial to achieve low carbon energy transitions. Low carbon innovation is a complex process, which includes R&D, demonstration, diffusion and deployment (Watson, 2008). It is not only a technological process but also depends on the existence of appropriate incentives for firms and other organisations to engage in technology development and on the creation of markets for technologies that have been successfully commercialised. Low carbon innovation involves a range of actors including firms, public bodies and institutions and research organisations. Innovation is also relevant for fossil fuel technology; however, as fossil fuels have been established technologies for many decades, innovation is less rapid and less groundbreaking in this field.

Innovation is here defined as creating something new, developing a new product, service or idea. According to Everett Rogers, innovation is usually perceived as uncertain and risky. It requires a group of individuals with the same interests to first adopt a technology and then spread the word. This process usually takes months or years before the diffusion of innovation takes place (Rogers, 2003). Innovation can be new to the world, new to the market or new to a specific setting (e.g. a country or a value chain). It does not only involve technologies or a product, but can also be innovation in business models, deployment models or service.

The various stages of technological development from R&D to commercialisation include early R&D, early demonstration (both usually covered by public funding), refinement and cost reduction, early commercialisation and fully commercialised (the last two usually covered by private funding). The most difficult stage, or so-called valley of death, is between the early demonstration stage and the early commercialisation, as public funding will often end here and private funding may not be available yet (Ockwell & Mallett, 2013).

Brewer (2007a) refers to the need to facilitate innovation and diffusion through international technology transfer – North to South, South to North and South to South – under the technology transfer paradigm.

Access to energy technology: technology transfer and cooperation

Many poor countries are suffering from a too low energy generation capacity, which is too low quantity and quality and inadequately priced for meeting the energy needs of their people. This means that energy security is weak, while at the same time energy access is low and energy poverty is high. The lack of modern energy access is therefore a constraint for growing the economies of poor countries and increasing the living standards and well-being of poor people. One of the major problems associated with this dilemma is that many of these countries do not have access to modern energy technology. This is why technology transfer and technology cooperation plays an important role.

The Intergovernmental Panel on Climate Change (IPCC)'s Special Report on Methodological and Technological Issues in Technology Transfer defines technology transfer as a:

> broad set of processes covering the flows of know-how, experience and equipment for mitigating and adapting to climate change. . . . The broad and inclusive term 'transfer' encompasses diffusion of technologies and technology cooperation across and within countries. It comprises the process of learning to understand, utilise and replicate the technology, including the capacity to choose it and adapt it to local conditions and integrate it with indigenous technologies.
>
> *IPCC (2000: 55)*

This definition thus includes both technology transfer and technology cooperation. Technology transfer describes a flow of know-how, experience and equipment from one firm or country to another (often from North to South such as from the EU to sub-Saharan Africa), whereas technology cooperation describes a flow of information from one firm or country to another in joint forms, such as in joint ventures, joint R&D, staff exchanges, licencing agreements, mergers and acquisitions, etc.

In the past, technology transfer and cooperation mainly focussed on the transfer of hardware, such as power plants or wind turbines, while neglecting the transfer of skills, knowledge and expertise, such as for how to run, manage and maintain a power plant or a wind turbine. In recent years, this understanding has been broadened. There are three flows within the process of technology transfer: (a) capital goods and equipment, (b) skills and know-how for operation and maintenance, and (c) knowledge and expertise for innovation (know why) (Ockwell et al., 2007; Ockwell & Mallett, 2012). Technology transfer and cooperation is thus not only limited to the hardware but also to the software, such as skills and knowledge. There is also a difference between horizontal and vertical technology transfer and cooperation. Technology transfer and collaboration between firms is called horizontal (e.g. from firm A to firm B) and advances in technology development refer to vertical technology transfer and collaboration (e.g. from R&D to commercialisation of a technology).

Technology transfer and technology cooperation, however, not only require access to the technologies, but also require innovation capacity for those who innovate and transfer the technology and absorptive capacity for those who receive the capacity.

The term innovation capacity (also referred to as innovative capacity) describes a firm's or country's ability to innovate. This requires having the skills, knowledge, expertise, technological infrastructure (such as laboratories and factories), financial means, institutional and logistical capacity, supportive policies and regulations, and government support in place to be able to innovate. Sustaining absorptive and innovative capacity requires a multi-faceted enabling environment (IPCC, 2000).

The term absorptive capacity describes a recipient firm's or country's ability to absorb new technology to enable the recipient to move from technology transfer to innovation and the development of indigenous markets (Ockwell et al., 2007; Ockwell & Mallett, 2012). At the same time, absorptive capacity is important for being able to run, manage and maintain the technologies and the impact the technologies have. This may be less of a problem for small-scale technologies, such as PV systems (although the recipients of the technology will need to know how to install, maintain and repair the PV system and its components), but it is certainly an issue for large-scale energy technology such as large hydropower dams. Recipient countries and firms need to have adequate procedures and policies in place with regard to the environmental and social impacts of dams (e.g. resettlement of local people, compensation, wildlife rescue plans, knowledge about how to conduct a proper EIA), how to operate and maintain the dam (how the water is being released and controlled, how turbines can be opened and shut, dam safety, repair works, etc.). This requires the thorough and long-term training of engineers, workers, policymakers and support staff not only to receive the technology but also to operate and manage it successfully in the future (Urban et al., 2015).

Indigenous innovation refers to the ability of firms or countries to receive technology, create new domestic production capacity and accumulate sufficient innovation capacity to be able to innovate independently. This means that technology transfer is not required any longer, as sufficient domestic capacity for innovation exists. Often indigenous innovation is not developed directly after technology transfer, but may be the result of a series of technology cooperations. With regard to low carbon energy technology innovation, China, for example, has developed excellent indigenous innovation for hydropower dams and solar water heaters. For wind energy, technology transfer used to be the main form of getting access to reliable wind turbines, followed by technology cooperation such as licencing for learning through 'reverse engineering' (de-mounting the turbines into their single components to understand how they were engineered and then aiming to manufacture the same turbine based on this knowledge). This has been followed by a phase of mergers and acquisitions, most importantly with the Chinese wind firm Goldwind purchasing the German R&D bureau Vensys. Chinese engineers now have access to Vensys technologies and are engaging in what they call indigenous innovation, which includes adapting some of the turbine models to Chinese conditions, such as low wind speeds, high altitudes, desert conditions, etc. (Urban, 2018). That said, other countries in the OECD might have a stricter understanding of what indigenous innovation is and may refer to it exclusively as developing new products or services independently of other firms or countries.

A schematic overview of technology transfer is shown in Figure 15.6.

Ockwell et al. (2015) suggest that, from a poverty reduction perspective, it is more beneficial to invest in technology transfer of small-scale renewable energy, rather than large-scale renewable energy technology, such as through the Clean Development Mechanism (CDM). This follows the logic that small-scale renewable

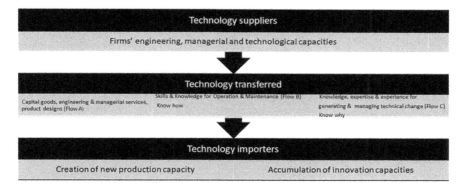

FIGURE 15.6   Schematic overview of technology transfer

Source: Adapted from Ockwell and Mallett (2013: 116)

energy technology helps local people to overcome energy poverty by providing energy access and enables the training of local engineers, technicians and users of the technology allowing them to manage, maintain and repair the energy technology over its lifetime. In contrast, the CDM frequently leads to investment in hardware rather than a transfer of knowledge, skills and expertise, and often does not benefit the local population or contribute to energy access. For example, some large hydropower dams that have been built under the CDM supply electricity to big cities or industrial development rather than contributing to alleviating energy poverty for the local population that is affected by the dam. In this case, small hydropower plants, such as river run-offs, micro-wind turbines or solar PV, might be more beneficial for the local population and their energy access needs.

Urban (2018) argue that emerging economies like China, India, Brazil and South Africa challenge the dominant North–South technology transfer paradigm. Increasingly it is being observed that countries such as China have increased not only their production capabilities but also their innovation capabilities, making them leaders in renewable energy technology such as in hydropower or solar PV. Similar developments can be seen in Brazil, which is a leader for sugar cane-based bioethanol for the transport sector.

Intellectual property rights

There is a conflict of interest between diffusion and development. Technology transfer is one of the issues which makes the divide between industrialised and developing countries particularly apparent, as both have different needs: developing countries primarily wish to gain access to clean technology and intellectual property rights (IPRs) for their development and economic growth, while industrialised countries are primarily interested in diffusion of their technologies, accessing commercial markets and protecting their IPRs (Ockwell et al., 2010).

IPRs are legally recognised rights that protect the author or originator of ideas, creations, products and services, for example, in the domains of technology, inventions, discoveries, designs, publications, music and artwork. IPRs include patents, trademarks, copyrights, industrial design rights and sometimes even trade secrets.

IPRs may be the most fiercely discussed issue in the technology transfer debate. IPRs divide the main players into two groups: those who want to strengthen the protection of IPRs and those who want to loosen it. On the one hand, it is argued that developing countries need to tighten their legal frameworks increasing IPR protection to cut out cheap imitations; on the other, it is argued that low carbon technology should be a public good and that IPR protection needs to be loosened to facilitate access to modern low carbon, climate-resilient technology to poor countries (Ockwell et al., 2010; UNFCCC, 2011). Scientifically, there is high uncertainty both ways: It is not clear whether IPRs are a catalyst or a barrier to technology transfer, since the different stages of technology development are poorly understood (Ockwell et al., 2010). Industrialised countries hold the majority of patents, while developing countries – and particularly low-income countries – hold a marginal share of patents (CAFOD, 2001). This is problematic, as poor countries need access to modern energy technology to reduce energy poverty and to follow a low carbon, climate-resilient development pathway. The process of registering patents is complicated, lengthy and expensive. This seems to be a major reason that many small- and medium-sized enterprises (SMEs) and companies in developing countries do not register their patents.

Under the UNFCCC, new mechanisms are being discussed to deal with IPRs for climate-relevant technologies. This could include the pooling of patents for use in developing countries at low-cost rates. It could also incorporate low-cost or free access to publicly funded technologies and concessional rates for privately funded technologies in accordance with the current development status of countries. Intellectual property rules may need to be shortened and facilitated to make the process easier for low-income countries with regard to patenting. At the same time, developing countries need to respect IPRs. Low-income countries could have privileged access and privileged pricing for intellectual property (IP), enabling them to access climate-relevant technologies both for adaptation and mitigation.

The problem is that many of the IPRs for climate-relevant technologies, particularly energy technology, are held by private firms. Understandably, the main interests of private firms are profits, protecting their IPRs, keeping their market shares and remaining competitive; and in return they guarantee employment and tax revenue for the state. Even if states sign up to reformed IPR rules for the sake of climate change, they will only be able to provide improved access to publicly held IPRs, rather than those that are privately held.

At the same time, access to IPRs does not guarantee that a country has the absorptive capacity to absorb, operate, maintain and manage the technology or the innovation capacity to engineer and build the technology independently. Access to IPRs is therefore only a small part of the technology transfer debate, albeit a strongly debated one.

Exercises

1   Make a list of the latest state-of-the-art technology for energy efficiency and low carbon energy. Briefly explain how each of these technologies work, as well as their advantages and disadvantages.
2   Imagine you are a policymaker, representative of a donor organisation or a representative of a firm. You are involved in the technology transfer and cooperation of renewable energy technology from one country to another. What do you need to consider? What are the opportunities and barriers? Which steps would you take to ensure the transfer of technology and the transfer of knowledge (both 'know how' and 'know why') will be successful? You may wish to prepare a short PowerPoint presentation to present your case study.

References

Atherton, K.D. (18 February 2013) Fusion power could happen sooner than you think. *Popular Science*. Available from: www.popsci.com/technology/article/2013-02/fusion-power-could-happen-sooner-you-think

Brewer, T.L. (2007a) *Technology, Emissions Trading and International Trade*. New Issues and Paradigms for the Expanding Climate Change Agenda. Annual Meeting of the Swedish Network of European Economists, Mölle.

CAFOD. (2001) *Intellectual Property Rights and Development*. London, CAFOD.

Caracappa, F. (2011) *Fukushima Accident: Radioactive Releases and Potential Dose Consequence*. Special Session: The Accident at Fukushima Daiichi – Preliminary Investigations, ANS Annual Meeting, 28 June 2011. Available from: www.ans.org/misc/FukushimaSpecial Session-Caracappa.pdf

Casillas, C. & Kammen, D. (2010) The energy–poverty–climate nexus. *Science*, 26 (330), 1181–1182.

Dequech-Neto, N. & Heiss, E. (2013) From outsider to world leader: Bioethanol in Brazil. In: Urban, F. & Nordensvard, J. (Eds.) *Low Carbon Development: Key Issues*, Earthscan. pp. 284–296.

Desertec Foundation (2018) *Desertec*. Available from: www.desertec.org/

Diesendorf, M. (2014) *Sustainable Energy Solutions for Climate Change*. Oxon, Earthscan, Routledge.

Enercon (2007) The world's most powerful wind turbine installed near Emden. *Windblatt*, 04/2007, 4. Available from: www.enercon.de/fileadmin/Redakteur/Medien-Portal/windblatt/pdf/en/WB-0407-en.pdf

ESHA. (2009) *SHP in the World. Small Hydro Power*, European Small Hydropower Association (ESHA).

Evans, A., Strezov, V. & Evans, T.J. (2008) Assessment of sustainability indicators for renewable energy technologies. *Renewable and Sustainable Energy Reviews*, 15 (3), 1082–1088.

Fridahl, M and Lehtveer, 2018. Bioenergy with carbon capture and storage (BECCS): Global potential, investment preferences, and deployment barriers. *Energy Research & Social Science,* 42 (8), 155–165.

Gagnon, L., Belanger, C. & Uchiyama, Y. (2002) Life-cycle assessment of electricity generation options: The status of research in year 2001. *Energy Policy*, 30, 1267–1278.

Global Solar Council. (2018). *Global Market Outlook 2019–2023*. Available from: http://www.solarpowereurope.org/global-market-outlook-2019-2023/

Global Wind Energy Council (GWEC). (2019) *Global Wind Report 2018*. Available from: https://gwec.net/wp-content/uploads/2019/04/GWEC-Global-Wind-Report-2018.pdf

Goldemberg, J. & Lucon, O. (2009) *Energy, Environment and Development*. 2nd edition. Oxon, Earthscan, Routledge.

IEA. (2019) *Statistics*. Paris, International Energy Agency (IEA), OECD/IEA. Available from: www.iea.org/statistics/

International Hydropower Association (IHA). (2019) *2019 Hydropower Status Report*. Available from: https://www.hydropower.org/statusreport

IPCC. (2000) Enabling environments for technology transfer. [Lead Authors: McKenzie Hedger, M., Martinot, E. & Onchan, T.] In: *Methodological and Technological Issues in Technology Transfer*. A Special Report of IPCC Working Group III of the Intergovernmental Panel on Climate Change IPCC. [Bert Metz, Ogunlade R. Davidson, Jan-Willem Martens, Sascha N.M. van Rooijen and Laura Van Wie McGrory (Eds.)]. Cambridge, Cambridge University Press. pp. 105–141.

Kato, E., Moriyama, R. & Kurosawa, A. (2017) A sustainable pathway of bioenergy with carbon capture and storage deployment. *Energy Procedia*, 114 (7), 6115–6123.

Lema, R., Nordensvärd, J., Urban, F. & Lutkenhorst, W. (2014) *Technological Trajectories for Climate Change Mitigation in Europe, China and India*. GDI, Bonn. Wind Energy in Europe. GDI Working Paper.

Levihn, F. (2017) *Bio Energy Carbon Capture and Storage: möjligheter och potential i Stockholm*. Fortum Värme, Stockholm.

Mander, S., Anderson, K., Larkin, A., Gough, C. & Vaughan, N. (2017) The role of bio-energy with carbon capture and storage in meeting the climate mitigation challenge: A whole system perspective. *Energy Procedia*, 114 (7), 6036–6043.

McKinsey and Company (2010a) *Energy Efficiency: A Compelling Global Resource*. New York, McKinsey.

Millar, R.J., Fuglestvedt, J.S., Friedlingstein, P., Rogelj, J., Grubb, M.J., Matthews, H.D., Skeie, R.B., Forster, P.M., Frame, D.J. & Allen, M.R. (2017) Emission budgets and pathways consistent with limiting warming to 1.5 °C. *Nature Geoscience*, 10, 741–747.

Mintenig, J., Khabbazan, M.A. & Held, H. (2017) The role of bioenergy and carbon capture and storage (BECCS) in the case of delayed climate policy – insights from cost-risk analysis. *Earth System Dynamics*. https://doi.org/10.5194/esd-2017-117

Ockwell, D. & Mallett, A. (Eds.) (2012) *Low Carbon Technology Transfer: From Rhetoric to Reality*. Abingdon, Routledge.

Ockwell, D. & Mallett, A. (2013) Low carbon innovation and technology transfer. In: Urban, F. & Nordensvard, J. (Eds.) *Low Carbon Development: Key Issues*. Abingdon, Routledge.

Ockwell, D., Mallett, A., Haum, R. & Watson, J. (2010) Intellectual property rights and low carbon technology transfer: The two polarities of diffusion and development. *Global Environmental Change*, 20, 729–738.

Ockwell, D., Sagar, A. & de Coninck, H. (2015) Collaborative research and development (R&D) for climate technology transfer and uptake in developing countries: Towards a needs driven approach. *Climatic Change*, 131 (3), 401–415.

Ockwell, D., Watson, J., MacKerron, G., Pal, P., Yamin, F., Vasudevan, N. & Mohanty, P. (2007) *UK-India Collaboration to Identify the Barriers to the Transfer of Low Carbon Energy Technology*. London, DEFRA.

Pour, N., Webley, P.A. & Cook, P.J. (2017) A sustainability framework for bioenergy with carbon capture and storage (BECCS) technologies. *Energy Procedia*, 114 (7), 6044–6056.

Raftery, A.E., Zimmer, A., Frierson, D.M.W., Startz, R. & Liu, P. (2017) Less than 2 °C warming by 2100 unlikely. *Nature Climate Change*, 7, 637–641.

Rathmann, R., Szklo, A. & Schaeffer, R. (2010) Land use competition for production of food and liquid biofuels: An analysis of the arguments in the current debate. *Renewable Energy*, 35 (1), 14–22.

REN21. (2019) *Renewables 2019 – Global Status Report*. Renewable Policy Network for the 21st Century (REN21). Available from: http://ren21.net/gsr-2019/?gclid=EAIaIQ obChMIpaLy-t6P5AIVi5AYCh1GSAy8EAAYASAAEgKecfD_BwE

Rogers, E.M. (2003) *Diffusion of Innovations*. 5th edition. New York, Free Press.

Stockholm Exergi. (2019) *BECCS*. Available from: www.stockholmexergi.se/om-stockholm-exergi/miljo-och-hallbarhet/beccs/

Tawney, L. (2013) Revolutionizing the electricity sector: Renewable energy and low carbon grid infrastructure. In: Urban, F. & Nordensvärd, J. (Eds.) *Low Carbon Development: Key Issues*. Oxon, Earthscan, Routledge. pp. 151–162.

UNEP/Grid-Arendal (2009) *Energy Consumptions and CO_2 Emissions from Building*. Available from: www.grida.no/graphicslib/detail/energy-consumption-and-co2-emissions-from-building_9ab8

UNFCCC. (2011) *Technology Transfer Framework*. United Nations Framework Convention on Climate change (UNFCCC). Available from: http://unfccc.int/ttclear/templates/render_cms_page?TTF_ene

Urban, F. (2014) *Low Carbon Transitions for Developing Countries*. Oxon, Earthscan, Routledge.

Urban, F. (2018) China's rise: Challenging the North – South technology transfer paradigm for climate change mitigation and low carbon energy. *Energy Policy*, 113, 320–330.

Urban, F. & Mitchell, T. (2011) *Climate Change, Disasters and Electricity Generation. Strengthening Climate Resilience*. Discussion Paper 8. Available from: www.odi.org/sites/odi.org.uk/files/odi-assets/publications-opinion-files/7151.pdf

Urban, F., Siciliano, G., Sour, K., Lonn, P.D., Tan-Mullins, M. & Mang, G. (2015) South-south technology transfer of low carbon innovation: Chinese large hydropower dams in Cambodia. *Sustainable Development*, 23(7–8), 232–244.

US Department of Energy. (2017) *DOE Announces Major Milestone Reached for Illinois Industrial CCS Project*. Available from: www.energy.gov/fe/articles/doe-announces-major-milestone-reached-illinois-industrial-ccs-project

Vestas (2019) *The V164–8.0 MW Breaks World Record for Wind Energy Production*. MHIVestas Offshore Wind. Available from: www.mhivestasoffshore.com/v164-8-0-mw-breaks-world-record-for-wind-energy-production/

Von Hippel, F.N. (2011) The radiological and psychological consequences of the Fukushima Daiichi incident. *Bulletin of the Atomic Scientists*, 67 (5), 27–36.

Watson, J. (2008) *Setting Priorities in Energy Innovation Policy: Lessons for the UK*. Cambridge, MA, Belfer Center for Science and International Affairs, Kennedy School of Government, Harvard University. ETIP Discussion Paper.

WEC. (2007) *2007 Survey of Energy Resources*. World Energy Council (WEC). Available from: http://minihydro.rse-web.it/Documenti/WEC_2007%20Survey%20of%20Energy%20Resources.pdf

WEC. (2017) *Energy Resources*. World Energy Council (WEC). Available from: www.worldenergy.org/data/resources/

WHO. (2005) *Chernobyl: The True Scale of the Accident*. Geneva, World Health Organization (WHO). Joint News Release WHO/IAEA/UNDP. Available from: www.who.int/mediacentre/news/releases/2005/pr38/en/index.html

WNA. (2019) *Nuclear Power in the World Today.* World Nuclear Association (WNA). Available from: https://www.world-nuclear.org/information-library/current-and-future-generation/nuclear-power-in-the-world-today.aspx

World Wind Energy Association (WWEA). (2019). *Statistics.* Available from: https://wwindea.org/information-2/information/

16

POLICY RESPONSES TO ENERGY POVERTY

Energy and the UN's development goals

Energy and the Sustainable Development Goals

In 2015, the UN set new development goals, called the Sustainable Development Goals (SDGs). These replaced the former Millennium Development Goals (MDGs). The SDGs explicitly incorporate energy and development targets as well as climate targets, and are therefore building on the lessons learnt from the MDGs, which did not include energy and climate goals. The SDGs are embedded in the Rio+20 agenda on sustainable development, which followed the Rio+20 Earth Summit in 2012.

The SDGs are as follows:

Goal 1. End poverty in all its forms everywhere

Goal 2. End hunger, achieve food security and improved nutrition, and promote sustainable agriculture

Goal 3. Ensure healthy lives and promote well-being for all at all ages

Goal 4. Ensure inclusive and equitable quality education and promote life-long learning opportunities for all

Goal 5. Achieve gender equality and empower all women and girls

Goal 6. Ensure availability and sustainable management of water and sanitation for all

Goal 7. Ensure access to affordable, reliable, sustainable and modern energy for all

Goal 8. Promote sustained, inclusive and sustainable economic growth, full and productive employment and decent work for all

Goal 9. Build resilient infrastructure, promote inclusive and sustainable industrialization and foster innovation

Goal 10. Reduce inequality within and among countries

Goal 11. Make cities and human settlements inclusive, safe, resilient and sustainable

Goal 12. Ensure sustainable consumption and production patterns

Goal 13. Take urgent action to combat climate change and its impacts[1]

Goal 14. Conserve and sustainably use the oceans, seas and marine resources for sustainable development

Goal 15. Protect, restore and promote sustainable use of terrestrial ecosystems, sustainably manage forests, combat desertification, halt and reverse land degradation and halt biodiversity loss

Goal 16. Promote peaceful and inclusive societies for sustainable development, provide access to justice for all and build effective, accountable and inclusive institutions at all levels

Goal 17. Strengthen the means of implementation and revitalize the global partnership for sustainable development.

UNDESA (2014)

Goal 7 explicitly addresses energy and development issues. More precisely:

Goal 7. Ensure access to affordable, reliable, sustainable, and modern energy for all

7.1 by 2030 ensure universal access to affordable, reliable, and modern energy services

7.2 increase substantially the share of renewable energy in the global energy mix by 2030

7.3 double the global rate of improvement in energy efficiency by 2030

7.a by 2030 enhance international cooperation to facilitate access to clean energy research and technologies, including renewable energy, energy efficiency, and advanced and cleaner fossil fuel technologies, and promote investment in energy infrastructure and clean energy technologies

7.b by 2030 expand infrastructure and upgrade technology for supplying modern and sustainable energy services for all in developing countries, in particular least developed countries and small island developing states.

UNDESA (2014)

At the same time, other SDGs are cross-referenced to issues that are relevant for SDG 7 on energy access. SDG 12 on sustainable consumption and production calls for a phase-out of fossil fuel subsidies, and SDG 13 on climate change sets targets that have relevance for energy issues too.

While the SDGs are a big step forward compared with the MDGs with regard to energy and development issues, they still fall short on many issues. For instance, some targets remain vague, such as those related to technology transfer and cooperation. Some SDGs also contradict each other. For example, the climate change goal may be difficult to achieve if economic growth and industrialisation are being strongly promoted.

Indicators and progress so far for SDG 7 on sustainable energy

The following indicators are being measured to identify progress on SDG 7:

7.1.1 Proportion of population with access to electricity

7.1.2 Proportion of population with primary reliance on clean fuels and technology

7.2.1 Renewable energy share in the total final energy consumption

7.3.1 Energy intensity measured in terms of primary energy and GDP [gross domestic product]

7.A.1 Mobilized amount of US dollars per year starting in 2020 accountable towards the US$100 billion commitment of climate-relevant technology finance as outlined in the UNFCCC Paris Agreement

7.B.1 Investments in energy efficiency as a percentage of GDP and the amount of foreign direct investment in financial transfer for infrastructure and technology to sustainable development services

Source: UN (2017b)

In 2018 and 2019, the UN reported limited progress towards achieving SDG 7:

> The world is not currently on track to meet Sustainable Development Goal 7 (SDG7), which calls for ensuring 'access to affordable, reliable, sustainable and modern energy for all' by 2030. Current progress falls short on all four of the SDG7 targets, which encompass universal access to electricity as well as clean fuels and technologies for cooking, and call for a doubling of the rate of improvement of energy efficiency, plus a substantial increase in the share of renewables in the global energy mix.
>
> *(IRENA, 2019: 1)*

> While overall progress falls short on meeting all targets, real gains are being made in certain areas. Expansion of access to electricity in poorer countries has recently begun to accelerate, with progress overtaking population growth for the first time in sub-Saharan Africa. Energy efficiency continues to improve, driven by advances in the industrial sector. Renewable energy is making impressive gains in the electricity sector, although these are not being matched in transportation and heating – which together account for 80 percent of global energy consumption. Lagging furthest behind is access to clean cooking fuels and technologies – an area that has been typically overlooked by policymakers. Use of traditional cooking fuels and technologies among a large proportion of the world's population has serious and widespread negative health, environmental, climate and social impacts.
>
> *(IRENA, 2018: 1)*

Progress has been made in providing electricity access to the world's poor, with nearly 120 million people getting access to electricity every year. While this is a great improvement, this would mean that in 2030 more than 650 million people will still live in energy poverty (IRENA, 2019). Access to clean cooking fuels is roughly at about 60% globally, while about 2.7 billion people are still without access to clean cooking fuels. This figure has reduced from about 3 billion in 2012. The share of renewable energy in electricity generation grew slightly to nearly 25%, the large majority coming from hydropower (IEA, 2019). The UN further reports:

> Three quarters of the world's 20 largest energy-consuming countries had reduced their energy intensity – the ratio of energy used per unit of GDP. The reduction was driven mainly by greater efficiencies in the industry and transport sectors. However, that progress is still not sufficient to meet the target of doubling the global rate of improvement in energy efficiency.
>
> *(UN, 2017a: 9–10)*

This critical account raises four questions:

First, are the SDG 7 targets too ambitious or even unrealistic? Will it be possible to achieve universal energy access for everyone everywhere? How will a world that is plagued by poverty, hunger, violent conflict and war, terrorism, ill health and inequality make energy access a priority? Who will be accountable and who will pay for this?

Second, is the time frame for measuring these improvements too short? The goals should ultimately be achieved by 2030, yet the UN's reporting period so far (between 2012 and 2019) was only a few years, hence is it too short for a real change?

Third, how is modern energy access being measured? Are all people accounted for? Or are definitions and measurements ill defined? The Indian government recently reported that 100% village electrification has been achieved, yet this does not mean that everyone has access to electricity, as the definition of electrification and the ways of measuring this are flawed. In the Indian case, only 10% of the households in a village and at least one communal building (like a school or clinic) has to have access to electricity to be classified as electrified (Indian Government, 2019). Yet, this can mean that 90% of the households in the village may not have access to electricity. Hence, any data on electrification will under-report the actual number of people not having access to electricity.

It has to be acknowledged, though, that some progress has been made so far on all levels, be it with regards to improvements in energy efficiency, increasing the share of renewable energy, or increasing access to electricity and clean cooking fuels. Improvements in electricity access have particularly been made in Asia, where electrification rates have increased substantially for most countries within the last few years (IEA, 2019). One measurable variable in SDG 7 is the doubled rate of energy efficiency, which is ambitious. Even more ambitious is achieving 100% electricity access and modern cooking fuel access, given the current state of the world. Despite these problems, it has to be acknowledged that progress is happening, albeit at a slow rate. There is still about a decade left until 2030, so more progress is likely. Whether the progress will be sufficient is another issue.

Mechanisms for overcoming energy poverty

The UN's Sustainable Energy for All Initiative (SE4All)

A total of about 1 billion people worldwide do not have access to electricity and 2.7 billion people rely on traditional biomass, such as fuelwood and dung, for basic needs, such as cooking and heating. The majority live in sub-Saharan Africa and developing Asia (IEA, 2019). The UN made 2012 the International Year of Sustainable Energy for All and 2014–2024 the Decade of Sustainable Energy for All, and it has set a target for providing universal modern energy access by 2030. This initiative is called Sustainable Energy for All (SE4All) and involves governments around the world, the private sector and civil society. This target aims to provide access to electricity and clean cooking facilities to everyone in every country. This goal is linked to renewable energy provision, as the UN estimates that about two thirds of the rural population in developing countries will get access to electricity through decentralised renewable energy, such as wind, solar and small hydro. This will be delivered by renewable energy-powered mini-grids and off-grid solutions. Decentralised renewable energy, such as biogas, also plays a key role for providing the rural poor with access to clean cooking facilities (IEA, 2010).

The objectives of the SE4All initiative are as follows (SE4All, 2018a):

- to ensure universal access to modern energy services
- to double the global rate of improvement in energy efficiency
- to double the share of renewable energy in the global energy mix

You can see that these targets have been incorporated in the SDGs, albeit in amended form. The SE4All initiative lists 20 key target countries with prevailing energy poverty that are being targeted specifically through the initiative in terms of electrification, such as in Sub-Saharan Africa: Angola, Burkina Faso, Chad, Democratic Republic of Congo (DRC), Ethiopia, Kenya, Madagascar, Malawi, Mali, Mozambique, Niger, Nigeria, Sudan, South Sudan, Tanzania, Uganda and in Asia: Bangladesh, India, Myanmar, North Korea. In terms of access to clean cooking fuels, the following 20 countries are prioritised: Democratic Republic of Congo (DRC), Ethiopia, Ghana, Kenya, Madagascar, Mozambique, Nigeria, Sudan, Tanzania, Uganda and in Asia: Afghanistan, Bangladesh, China, India, Indonesia, Myanmar, North Korea, Pakistan, Philippines and Vietnam (SE4All, 2017).

Less than US$50 billion per year (roughly US$48 billion per year or US$1 trillion in total) is needed for universal energy access (IEA, 2011). Small-scale renewable and energy-efficient technologies are important for providing electricity generation and cooking in poor countries to achieve the SE4All target.

The SE4All targets also stress the role of renewable energy and energy efficiency. It is reported that investments in renewable energy have created 2.3 million jobs around the world. The SE4All initiative mentions several energy-efficient options that could save up to 50% of global emissions (SE4All, 2018b). Despite these positive outlooks, only time will tell if the targets will be achieved or whether in a few decades new initiatives and programmes will follow with new targets, such as has been the case for the MDGs that were superseded by the SDGs.

Other multilateral initiatives

This section discusses several multilateral initiatives for increasing modern energy access. This not only relates to straightforward energy access initiatives such as the SE4All initiative, but also relates more broadly to a set of initiatives that aim to increase access to modern energy technology and innovation in countries where energy poverty is prevalent. We will specifically discuss the contribution that the Clean Development Mechanism (CDM) makes to increasing poor countries' access to modern energy technology, as well as the role played by Climate Innovation Centres and the Technology Mechanism under the UNFCCC.

The Clean Development Mechanism

The CDM enables industrialised countries to finance projects leading to emission reductions in developing countries with the aim of contributing to sustainable development. Projects must be 'additional', meaning that investments should only be made in projects that create an additional greenhouse gas (GHG) emission reduction, rather than funding already planned or proposed projects. Developing countries gain access to climate-friendly technology, such as renewable energy, while industrialised countries gain certified emission reduction credits (CERs) to offset their emissions. This is how the CDM is linked to providing access to modern energy technology to poor countries. The CDM has mainly contributed to wind energy, hydropower and biomass energy projects in developing countries, less in other forms of climate-friendly technology. In March 2019, nearly 8,000 projects had been registered with the UNFCCC and another 580 were pre-registered. It is estimated that the current CDM project will account for over 1,028,800,000 CERs (ton of CO_2 equivalents) by 2020 (UNEP, 2019). The large majority of CDM projects are located in China, India and other emerging economies, whereas poorer countries have failed to attract many CDM projects. Over 80% of all CDM projects are located in Asia, about half of them in China (UNEP, 2019). Only about 1.5% of all CDM projects are located in least developed countries and only about 0.5% in small island developing states. More details about the CDM can be found in Chapter 13.

A key criticism of the CDM is that it is an offsetting mechanism, meaning that a country or a firm invests in low emissions projects in poor countries, rather than reducing their own emissions. Urban and Nordensvärd (2013) argue that this could be considered a modern form of selling of indulgences. Another criticism of the CDM relates to its inadequate contribution to sustainable development and limited additionally. Additionality means, in this context, that the projects should be built in addition to and going beyond existing plans (e.g. Karp & Liu, 2000; Ockwell et al., 2008; Boyd et al., 2009; Nussbaumer, 2009; Drupp, 2011). Higher standards, such as the Gold Standard, can improve this situation by aiming to ensure environmental, social and economic sustainability; however, the criticism is still that sustainability is often limited in CDM projects (Drupp, 2011). In terms of additionally, critics have raised concern that some projects would have been built anyway, such as some large hydropower dams, even if they had not received investments through the CDM.

There is ongoing debate as to whether the CDM has contributed to technology transfer and knowledge transfer; however, the overwhelming conclusion is that this has been limited and that the main transfer has been in financial form, in terms of generating investments in the recipient countries (Urban, 2014). While the aim of the CDM was to promote the North–South transfer of climate-friendly technologies, it has emerged that in some cases, such as for India, the actual transfer of technologies is very low. Dechezlepretre et al. (2009) calculated that only about 12% of CDM projects delivered an actual transfer of technology in India. This percentage is higher in other countries, although also rather low for Brazil. This means that the CDM has limited capabilities in providing access to modern energy technology for poor countries. In addition, the CDM contributes mainly to wind energy and hydropower projects (including large dams) in developing countries, less to other forms of climate-friendly technology. We have discussed the advantages and disadvantages of large dams throughout this book and while they can provide large amounts of electricity for poor countries, the electricity often tends to be transported elsewhere (e.g. to urban centres, industrial clusters or neighbouring countries) rather than being used by people living in energy poverty (usually the rural poor).

Byrne et al. (2014) argue for smaller-scale, pro-poor, climate-friendly energy technology transfer that builds up the indigenous innovative capacity of the recipient country as opposed to the large-scale, top-down initiatives of the CDM that mainly generate access to hardware and financial transactions. In the long run, it is important to build up local capacities and knowledge for it to be of value to people in low- and middle-income countries who can use these technologies on a daily basis to reduce energy poverty.

The Climate Innovation Centres (CICs)

Several regional Climate Innovation Centres were launched in recent years in a number of developing countries with the aim of increasing R&D, testing and diffusion of climate-relevant innovation, for both mitigation and adaptation. The centres are funded by InfoDev, a multi-donor trust fund administered by the World Bank, which aims to promote innovation and entrepreneurship in developing countries. The centres are part of the wider Climate Technology Program (CTP), which aims to build up the innovative capacity of developing countries for climate-friendly technology. CTP is funded by the Australian Agency for International Development (AusAID), the Danish International Development Agency (DANIDA), the UK's Department for International Development (DFID), the government of Norway and the World Bank. Climate Innovation Centres currently operate in the Caribbean (located in Jamaica), India, Ethiopia, Ghana, Kenya (the first one to be launched), Morocco, Nigeria, South Africa and Vietnam. They provide grants to local innovators and entrepreneurs to develop climate-friendly technology.

The Climate Innovation Centres thereby help countries and regions in which energy poverty is prevalent, such as in India and sub-Saharan Africa, to build up their indigenous capacity to innovate and develop their own modern energy technology. This is likely to have positive effects on increasing availability, affordability and access to modern energy technology for poor people.

The Technology Mechanism

The Technology Mechanism under the UNFCCC was set up at the international climate conference COP 16 in Cancun in 2010. It includes setting up the Climate Technology Centre and Network (CTCN). This aims to 'stimulate technology cooperation and to enhance technology development and transfer' between Annex I countries (developed countries) and non-Annex I countries (developing countries) (WEDO, 2014; UNFCCC, 2014). It helps to identify technology needs; for example, using the Technology Needs Assessment (TNA) that aims to address the technological needs developing countries have with regard to climate change adaptation and mitigation. 'It also facilitates the preparation and implementation of technology projects and strategies that support action on climate change' (UNFCCC, 2014: 1). The official host of the Technology Mechanism is the United Nations Environment Programme (UNEP). It offers three support services:

1 Providing technical assistance at the request of developing countries to accelerate the transfer of climate technologies
2 Creating access to information and knowledge on climate technologies
3 Fostering collaboration among climate technology stakeholders via the Centre's network of regional and sectoral experts from academia, the private sector, and public and research institutions.

UNEP (2018)

This can help to overcome energy poverty by proving access to climate-friendly energy technology.

Figure 16.1 shows how the Technology Mechanism operates.

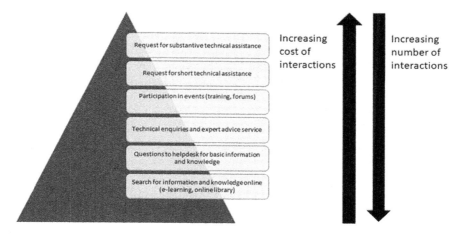

FIGURE 16.1 Overview of the Technology Mechanism established by the UNFCCC

Source: Based on CTCN (2018)

National initiatives

There are various national initiatives for overcoming energy poverty. This is highly context specific and depends on national policy, national funding and national priorities. However, many countries have rural electrification schemes in place to increase access to modern energy, mostly electricity, in rural areas. In many countries, rural electrification is achieved through renewable energy, for example, by using hydropower, such as in China several decades ago and across the Greater Mekong Sub-Region today, or with solar and wind energy, such as in Costa Rica, India, South Africa and Vietnam. It is often middle-income countries that have had effective rural electrification schemes, whereas low-income countries tend to lack the investment, engineering skills, technical know-how (and sometimes also the political capacity) to achieve large-scale electrification. While rural electrification has helped millions of people to get access to modern energy, some argue that rural electrification alone will not overcome energy poverty (e.g. Bhattacharyya, 2006a, 2006b). This is due to the low share of electricity used by the rural poor, who predominantly need modern energy, such as liquefied petroleum gas (LPG), natural gas, biogas and solar cookers, for cooking. This situation is unchanged by electrification schemes, which target other end-uses. Tackling energy poverty with regard to improved cooking stoves and modern cooking fuels may therefore have a greater impact on those living in energy poverty than focussing on electricity access only.

Other national initiatives to increase modern energy access and reduce energy poverty include microcredit schemes. Several international, national, regional and local banks, lenders and even some NGOs offer microcredit to people who are in financial need. Sometimes microcredit is non-conditional (meaning that the loan can be invested as desired by the recipient) or sometimes it is attached to specific conditions, such as being conditional on the purchase of agricultural or livestock produce or renewable energy technology. For example, Grameen Shakti in Bangladesh is a social enterprise that focusses on bringing renewable energy, mainly solar photovoltaic (PV) energy, to rural areas. It lends money to the rural poor to purchase solar energy technology. The aim is both to provide electricity to the energy poor and to promote income-generating activities. So far, this has led to solar-powered cafes, restaurants and tea stalls in the rural areas, solar-powered mobile phone-charging businesses, solar-powered boats, solar-powered water purification systems, etc. Often the customers who receive the microcredit are women. Grameen Shakti is a part of the Bangladeshi Grameen Bank, which has received several prizes for its work, including the Right Livelihood Award and two Ashden Awards. It is estimated that the organisation has installed more than 1 million solar PV systems in rural areas amounting to a combined capacity of more than 25 MW. The organisation also provides improved cooking stoves and biogas plants, available through microcredit.

The final initiative discussed here is related to both enabling modern energy and modern communication. Solar-powered telecommunications, such as PV-powered mobile phone masts, are being used by mobile phone operators in several African countries. For example, in Tanzania the telecommunications firm Celtel powers its mobile phone masts with solar energy. It provides electricity to power the antennae of the phone masts so that they can transmit radio signals between base stations and mobile phones. This is a new and innovative way of providing both modern telecommunications and energy. While this initiative does not directly reduce energy poverty, it helps to bring modern amenities and infrastructure to poorer countries.

An outlook

This book explored the main issues around energy and development. Access to energy is a key development issue and is a prerequisite for achieving wider development goals. At the same time, energy use is closely intertwined with environmental challenges such as climate change, fossil fuel resource depletion, air pollution and natural resource management (land, water, forests).

This book discussed the key issues and concepts in the field of energy and development. It addressed policy responses, such as the energy targets of the SDGs and the UN's target of universal energy access. The book further outlined various options for delivering energy access (both low carbon and fossil fuel based), their environmental, socio-economic and technological implications, and how these link to contemporary global challenges in the fields of environmental management and sustainable development.

Throughout the chapters, we have discussed how and why energy and development is a fiercely debated topic that is receiving increasing attention due to climate change, natural resource scarcity, prevailing global poverty and policy responses to these issues at the international, regional and national levels.

In terms of an outlook for the future, despite several decades of sustained efforts to reduce energy poverty, an estimated 1 billion people worldwide do not have access to electricity and 2.7 billion people rely on traditional biomass for basic needs such as cooking and heating (World Bank, 2019). The UN hopes that its universal energy access programme will eliminate energy poverty by 2030 and thereby provide every person with modern energy, such as electricity and clean cooking fuels. This is a grand and ambitious plan, considering the scale, global extent and complexity of the issue. Pessimists will say it is impossible to provide modern energy access to everyone worldwide. Pessimists will refer to the huge investment sums needed to achieve this, the limited time frame, and the occurrence of war, violent conflict and natural disasters (often exacerbated by climate change) that destroy existing amenities and services or that eradicate the hope for even the most basic energy facilities. Optimists will say it is possible to achieve the UN's universal energy access target by 2030 as we still have about a decade of technological achievement, sustained investment, global commitment and relentless effort ahead of us. Technocrats will most likely emphasise the technical opportunities and the latest energy innovations that exist today, making the energy access target technically feasible, while economists will stress its large-scale investment challenges.

Without being too negative or too positive, the truth probably lies somewhere in between. It is likely that the UN, other multilateral institutions and national governments will make progress over the next few decades with regard to providing energy access. It is likely that small-scale renewable energy technology, particularly solar PV, will play a major role in providing access to modern energy for rural households that cannot be reached by the grid under current geographic, technical and economic circumstances. It is also likely that clean cooking fuels, such as biogas, as well as improved cooking stoves, will reach more people than in the past and thereby improve people's lives, particularly those of women and children. Some technical innovations, such as solar-powered irrigation, solar-powered healthcare services and telecommunications can enable both energy access and improve overall living standards.

In terms of energy technology and innovation, there has probably never been a better opportunity than today to tackle energy poverty. Major investment will be needed around the globe; however, this could be funded to some extent from reducing (and eventually phasing out) fossil fuel subsidies, and hence a global low carbon energy access target is less expensive than it might at first appear. Renewable energy technology, particularly solar PV, has also seen rapid decreases in the levelised costs of electricity (LCOE) in recent years and most renewables have achieved grid parity in the last few years.

However, there are major challenges. Population growth is expected to counteract some of the progress that is being made in providing energy access, as more people will need to be lifted out of energy poverty. The UN predicts significant progress in providing modern energy access, but in absolute terms these efforts will be weakened by the strong population growth in many parts of the world (e.g. the world population was about 6.5 billion people in 2005, reached 7 billion in 2012 and is close to 8 billion in 2018, so more people need access to electricity and clean cooking fuels) (World Bank, 2019). This is particularly relevant for hotspots of energy poverty, such as India and sub-Saharan Africa that are also experiencing rapid population growth. At the same time, the vulnerability of some developing countries and the wide range of complex poverty issues means that providing energy access may not be at the top of a country's political agenda as there may be more pressing issues, such as providing clean water, increasing food security and reducing hunger, improving health, providing access to education, peace building and fighting terrorism. In investment terms, poor countries will need help from donors to stem the large sums that are needed for providing energy access. This requires a sustained financial commitment by wealthy donor countries and institutions that goes beyond pledges and lip service.

Finally, with the renewed interest in tackling energy poverty, the global attention that is given to this issue and the embedding of energy and development targets in the SDGs, as well as the SE4All targets, there is a good chance that substantial progress will be made in the coming decades. Even though many people might still be without modern energy access by 2030, it is very likely that millions more people will have access to modern energy services at cost-effective prices, as it has become a global priority that is much better understood today than in the past.

Exercises

1   What progress has been made on eradicating energy poverty worldwide? What has been achieved and where is progress still lacking?
2   Look at the following report: *Tracking SDG7: The Energy Progress Report 2019* by the International Energy Agency, the International Renewable Energy Agency, the United Nations, the World Bank Group and the World Health Organization. Available from https://irena.org/publications/2019/May/Tracking-SDG7-The-Energy-Progress-Report-2019

   Choose a country of your choice and examine the progress this country has made since SDG 7 was established. See the figures in the report for a quick overview. What progress has been made in this specific country? What needs further improvement?
3   What do you expect for the future in terms of energy and development?

Note

1   Acknowledging that the UNFCCC [United Nations Framework Convention on Climate Change] is the primary international, intergovernmental forum for negotiating the global response to climate change.

References

Bhattacharyya, S.C. (2006a) Energy access problem of the poor in India: Is rural electrification a remedy? *Energy Policy*, 34 (18), 3387–3397.
Bhattacharyya, S.C. (2006b) Renewable energy and the poor: Niche or nexus? *Energy Policy*, 34 (6), 659–663.

Boyd, E., Hultman, N., Roberts, J.T., Corbera, E., Cole, J., Bozmoski, A., Ebeling, J., Tippman, R., Mann, P., Brown, K. & Liverman, D.M. (2009) Reforming the CDM for sustainable development: Lessons learned and policy futures. *Environmental Science & Policy*, 12 (7), 820–831.

Byrne, R., Ockwell, D., Urama, K., Ozor, N., Kirumba, E., Ely, A., Becker, S. & Gollwitzer, L. (2014) *Sustainable Energy for Whom? Governing Pro-poor, low Carbon Pathways to Development: Lessons from Solar PV in Kenya.* STEPS Centre, Brighton. Working Paper 61.

CTCN. (2018) *Introduction to the Climate Technology Centre & Network.* Climate Technology Centre and Network (CTCN). Available from: www.unep.org/climatechange/ctcn/Services/Introduction/tabid/771787/language/en-US/Default.aspx

Dechezlepretre, A., Glachant, M. & Meniere, Y. (2009) Technology transfer by CDM projects: A comparison of Brazil, China, India and Mexico. *Energy Policy*, 36 (4), 1273–1283.

Drupp, M.A. (2011) Does the Gold Standard label hold its promise in delivering high Sustainable Development benefits? A multi-criteria comparison of CDM projects. *Energy Policy*, 39 (3), 1213–1227.

IEA. (2010) *World Energy Outlook 2010. Energy Poverty: How to Make Modern Energy Access Universal?* Paris, International Energy Agency (IEA), OECD/IEA.

IEA. (2011) *World Energy Outlook 2011. Energy for All. Financing Access for the Poor.* Paris, International Energy Agency (IEA), OECD/IEA.

IEA. (2019) *Statistics.* Paris, International Energy Agency (IEA), OECD/IEA. Available from: www.iea.org/statistics/

Indian Government, DDUGJY Scheme of Govt. of India for Rural Electrification. (2019) *Definition of Electrified Village.* Available from: http://www.ddugjy.gov.in/portal/definition_electrified_village.jsp

IRENA. (2018) *Tracking SDG7: The Energy Progress Report (2018).* Available from: https://www.irena.org/publications/2018/May/Tracking-SDG7-The-Energy-Progress-Report

IRENA. (2019) *Tracking SDG7: The Energy Progress Report (2019).* Available from: https://www.irena.org/publications/2019/May/Tracking-SDG7-The-Energy-Progress-Report-2019

Karp, L. & Liu, X. (2000) *The Clean Development Mechanism and Its Controversies.* University of California, Berkeley. CUDARE Working Paper 903. Available from: http://escholarship.org/uc/item/9739314q

Nussbaumer, P. (2009) On the contribution of labelled Certified Emission Reductions to sustainable development: A multi-criteria evaluation of CDM projects. *Energy Policy*, 37, 91–101.

Ockwell, D.G., Watson, J., MacKerron, G., Pal, P. & Yamin, F. (2008) Key policy considerations for facilitating low carbon technology transfer to developing countries. *Energy Policy*, 36, 4104–4115.

SE4All. (2017) *Access to Energy.* https://www.seforall.org/goal-7-targets/access

SE4All (2018a) *About Us.* United Nations (UN). Available from: www.seforall.org/about-us

SE4All (2018b) *Renewable Energy.* United Nations (UN). Available from: www.se4all.org/about-us_our-ambition_renewable-energy

UN. (2017a) *Report of the Secretary-General, 'Progress towards the Sustainable Development Goals. E/2017/66'.* United Nations. Available from: https://unstats.un.org/sdgs/files/report/2017/secretary-general-sdg-report-2017 – EN.pdf

UN. (2017b) *Sustainable Development Goal 7. United Nations.* Available from: https://sustainabledevelopment.un.org/sdg7

UNDESA. (2014) *Outcome Document – Open Working Group Proposal for Sustainable Development Goals. Sustainable Development Knowledge Platform.* United Nations Department of Economic and Social Affairs (UNDESA). Available from: http://sustainabledevelopment.un.org/focussdgs.html

UNEP. (2018.) *Climate Technology Center & Network*. United Nations Environment Programme (UNEP). Available from: http://web.unep.org/climatechange/what-we-do/climate-technology-centre-and-network

UNEP (2019). *CDM Pipeline*. Available from: http://www.cdmpipeline.org/

UNFCCC. (2014) *The Technology Mechanism*. United Nations Framework Convention on Climate Change (UNFCCC). Available from: http://unfccc.int/ttclear/templates/render_cms_page?TEM_ctcn

Urban, F. (2014) *Low Carbon Transitions for Developing Countries*. Oxon, Earthscan, Routledge.

Urban, F. & Nordensvärd, J. (Eds.) (2013) *Low Carbon Development: Key Issues*. Oxon, Earthscan, Routledge.

WEDO (2014) *Climate Change Briefing on Gender, Mitigation and Technology*. Women's Environment and Development Organization (WEDO), August 18. Available from: http://wedo.org/climate-change-briefing-on-gender-mitigation-and-technology/

World Bank (2019) *Data*. Washington DC, The World Bank. Available from: http://data.worldbank.org/

INDEX

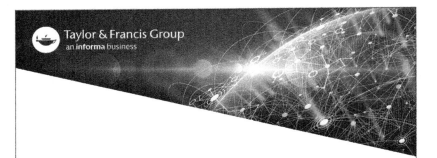